Y0-CAO-861

VMware NSX Micro-segmentation

Day 1

Wade Holmes, VCDX#15, CISSP, CCSK

Foreword by Tom Corn, Senior Vice President, VMware Security Products

vmware PRESS

VMWARE PRESS

Program Managers

Shinie Shaw
Katie Holms
Eva Leong

Technical Writer

Rob Greanias

Designer and Production Manager

Shirley Ng-Benitez
Wei-Pei Cherng

Warning & Disclaimer

Every effort has been made to make this book as complete and as accurate as possible, but no warranty or fitness is implied. The information provided is on an "as is" basis. The authors, VMware Press, VMware, and the publisher shall have neither liability nor responsibility to any person or entity with respect to any loss or damages arising from the information contained in this book.

The opinions expressed in this book belong to the author and are not necessarily those of VMware.

VMware, Inc. 3401 Hillview Avenue Palo Alto CA 94304 USA
Tel 877-486-9273 Fax 650-427-5001 www.vmware.com.

Copyright © 2017 VMware, Inc. All rights reserved. This product is protected by U.S. and international copyright and intellectual property laws. VMware products are covered by one or more patents listed at http://www.vmware.com/go/patents. VMware is a registered trademark or trademark of VMware, Inc. and its subsidiaries in the United States and/or other jurisdictions. All other marks and names mentioned herein may be trademarks of their respective companies.

Table of Contents

List of Figures

List of Tables

About the Author

Wade Holmes, VCDX#15, CISSP, CCSK, is a Senior Technical Product Manager within the VMware Networking and Security business unit and leads security architecture and solutions for the NSX Technical Product Management team. Wade has been with VMware for seven years, and has over 19 years of industry experience working on products and solutions within complex computing environments of all scopes and sizes. Wade's previously published work includes co-authoring the VMware vCloud® Architecture Toolkit, and numerous whitepapers and design guides. Wade was the first external VMware Certified Design Expert in the world, fifteenth overall, and is a VMware vExpert™.

Wade has presented at conferences such as VMworld US and Europe, Gartner Security Summit, SXSW Interactive, LISA, VMware Tech Summit US, VMware Tech Summit Tokyo, vForum Tokyo, multiple VMware User Group conferences, and has been an academic guest lecturer at the graduate level on security and Micro-segmentation. Wade holds a Bachelor's degree in Information Technology, a Master's degree in Information Assurance, and is pursuing a Doctoral degree in Cybersecurity.

You can contact Wade on Twitter @wholmes.

Content Contributors

Kausum Kumar is Senior Product Manager in the VMware Networking and Security business unit. Kausum has over 16 years of experience in the networking and security industry. Kausum leads the micro-segmentation and security area for VMware NSX with particular focus on firewalling, endpoint security and service chaining. Kausum has a Masters from University of Maryland, Baltimore County in Electrical Engineering with focus in wireless communications.

Stijn Vanveerdeghem (CCIE #36292) is a Senior Technical Product Manager in the VMware Networking and Security business unit. Stijn has been with VMware for over a year and has over 12 years of experience in the security field. Stijn's main areas of focus at VMware include NSX security features, security partners, and solving remote and branch office challenges with NSX.

Dale Coghlan is a Solution Architect in the VMware Networking and Security business unit and works directly with NSX for vSphere customers from initial design all the way through to implementation and operationalisation of their new environments. Dale has over 17 years of experience in networking and security roles across many verticals and uses that experience to help customers get the best out of the NSX network virtualization platform.

Additional Contributors:

Josh Batey
Anthony Burke
Hammad Alam
Dan Illson
Bruno Germain
Giles Chekroun

Acknowledgements

It takes the knowledge and resources of multiple individuals to create a technical book successfully. I would like to thank the following people for their support in developing and reviewing the material included.

Thank you to the VMware Networking and Security Business Unit Marketing team, including Shinie Shaw and Eva Leong (Program Managers) for your support in creating this book.

Thank you, Rob Greanias (Technical Writer) and Shirley Ng-Benitez (Designer and Production Manager) for your efforts in completing this book.

Thank you, Kausum Kumar, Stijn Vanveerdeghem, and other members of the NSX product team for your input, review, and support in this effort.

Thank you, Dale Coghlan, Josh Batey and Anthony Burke, and Hammad Alam of the VMware Networking and Security Business Unit Solutions Architect team for your review and input leveraging experience implementing NSX Micro-segmentation based security frameworks for organizations worldwide.

Thank you, Dan Illson, Bruno Germain, and Giles Chekroun of the VMware Networking and Security Business Unit Systems Engineering team for your input, review, and support.

Thank you to the VMware Networking and Security Business Unit management team including Tom Corn, Jeff Jennings, Dom Delfino, Milin Desai, Geoff Huang, Brian Lazear, Nikhil Kelshikar, and Jacob Rapp for your support and sponsorship in creating this book.

Wade Holmes

Wade Holmes, VCDX#15, CISSP, CCSK

Preface

VMware NSX Micro-segmentation - Day 1 offers guidance to security architects and practitioners planning to implement NSX for additional security and visibility through micro-segmentation. VMware NSX

Micro-segmentation - Day 1 provides the information needed to plan a security strategy around micro-segmentation using VMware NSX.

This includes guidance to plan, design, and implement security policy utilizing micro-segmentation.

Foreword

The security industry is at a crossroads. Growth in security spend is vastly outpacing growth in overall IT spend. The only thing growing faster than security spend, is security losses. This is more than simply a rapidly evolving threat landscape. Something more fundamental is at play.

Perimeter defenses can't stop every attack; the attack surface is simply too wide -- the perimeter too porous. And once inside, we lack the compartmentalization to contain attackers, and the visibility and control to detect and respond in kind. The problem is not a missing security product or feature, but rather an architectural gap -- a gap between what we're trying to protect (applications and data) and where we are trying to protect them from (servers and networks).

When applications were architected as monolithic stacks, the infrastructure and applications were aligned - and so were our controls. But when applications became distributed systems, that connection was broken. The comingling of all these distributed applications led to flat network architectures. When we try to implement security policy in these environments, we lack the handles to align those controls on the applications and data we're trying to protect. Instead we group together and segment assets with common attributes, such as web servers versus app or database servers, production versus development environments. And while there are plenty of logical policies to enforce on those boundaries, it comingles assets from many different applications, and does nothing to solve our most fundamental problems; lateral movement, alignment of controls, lack of context and actionability, and policy complexity. We are building skyscrapers on quick sand. We have all the right materials, but we lack the foundation or architecture for a sound system.

Enter micro-segmentation; Two years ago, VMware introduced a new approach that leverages network virtualization to segment environments around logical boundaries such as applications and regulatory scopes. In the two years since it has become one of the fastest growing categories in security. The reason goes back to its ability address these very issues - aligning controls and establishing least privilege environments around our critical applications. Doing so vastly decreases the attack surface, improves signal to noise, and greatly improves the actionability of all signals coming from our environments.

But moving to this kind of model requires change - change to how we design and operationalize segmentation. As with anything, there's a right way and wrong way to go about the exercise. Doing this correctly will yield incredible results of the protection of our assets in the simplification of our environments

It is so critical that practitioners become more educated on this approach. VMware NSX Micro-segmentation - Day 1 provides a simple and practical guide to get started. Wade brings clarity and presents a straightforward approach to secure architecture design through micro-segmentation. VMware invented this new approach and Wade and his peers have had an opportunity to work with large numbers of organizations going through this transition. This approach is based on the collective experience of the VMware field working with organizations implementing micro-segmentation around the world. The learning from that work has enabled him to create a book that distills that knowledge, and enables you to benefit from the learnings from our thousands of customers.

Security organizations are almost always the last to the party. Infrastructures are designed, applications are built, and then security teams are charged with securing them. Never before have we been able to define and implement a security architecture that makes sense - until now. Micro-segmentation and the use of network virtualization provides an opportunity to provide intrinsic security capabilities through a purpose built security architecture. We cannot miss this opportunity to move beyond the "bolt-on" model and truly architect-in security. VMware NSX Micro-segmentation - Day 1 will help you do just that.

Tom Corn, Senior Vice President,
VMware Security Products

Chapter 1

Introduction

The landscape of the modern data center is rapidly evolving. The migration from physical to virtualized workloads, the move towards software-defined data centers, the advent of a multi-cloud landscape, the proliferation of mobile devices accessing the corporate data center, and the adoption of new architectural and deployment models (e.g., micro-services and containers) are all driving a constant evolution. Users request higher levels of agility and service efficiency without compromise. However, this evolution is not without peril, as the impact on security often ends up being an afterthought. The operational dexterity achieved through the ability to rapidly deploy new applications has overtaken the ability of traditional networking and security controls to maintain an acceptable security posture. Additionally, there is a fundamental problem of traditionally structured security remaining inadequate even in conventional, static data centers.

Without a flexible approach to security and risk management, which adapts to the onset of new technology paradigms, security silos using disparate approaches are created. These silos act as control islands, making it difficult to apply risk-focused predictability within a corporate security posture. This allows unforeseen risks to be realized. These actualized risks cause an organization's attack surface to grow as the adoption of new compute technology increases, increasing susceptibility to advanced threat actors.

A foundational aspect of addressing this problem is the implementation of micro-segmentation. NSX is a networking and security platform able to deliver micro-segmentation across all the evolving components comprising the modern data center. NSX-based micro-segmentation increases the agility and efficiency of a data center while maintaining an acceptable security posture. VMware NSX Micro-segmentation - Day 1 details the security requirements needed to provide effective security controls and risk management within the modern data center, clearly showing how to go beyond the automation of legacy security paradigms by operationalizing agile security through NSX micro-segmentation.

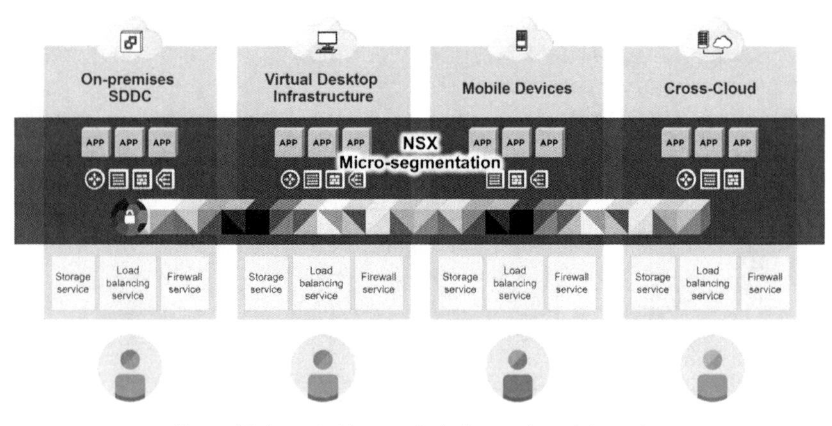

Figure 1.1 Acceptable security in the modern data center

It is no longer acceptable to utilize the traditional approach to data center network security - a legacy model built around a strong perimeter defense with minimal protection inside the perimeter. This model offers little protection against the most common and costly attacks targeting organizations today, including attack vectors originating within the perimeter. These attacks infiltrate the perimeter, learn the internal infrastructure, and laterally spread through the data center.

The ideal solution to complete data center protection is to protect every traffic flow inside the data center with a firewall, allowing only the flows required for applications to function. This is known as the Zero Trust Model. Achieving this level of protection and granularity with a traditional firewall is operationally unfeasible and cost prohibitive, as it would require traffic to be hair-pinned to a central firewall with individual virtual machines placed on distinct VLANs (i.e., Pools of Security model).

Without NSX, east-west protection of virtual machine requires implementation of the Pools of Security model. Significant network resource utilization bottlenecks are created by sending east-to-west communication from every VM to every other VM through a physical firewall. If capacity in a physical firewall is exhausted, there are only two ways to scale security in an environment; either replace with a larger firewall or add additional physical firewall, with the latter requiring major traffic re-engineering. If attempting to create a least-privilege model with physical firewalls, VLAN resources (limited to 4096) will quickly be exhausted by segmenting workloads into application-centric pools of security. The hairpinning of traffic through physical firewalls can also create additional latency for certain applications. These fundamental constraints of traditional perimeter-centric security architectures impact both security posture and application scalability within modern data centers.

An example of this scenario is shown in Figure 1.2 below. Without NSX, all application network traffic must traverse a physical firewall to be segmented, even when residing on the same physical server. With NSX, application network traffic can be efficiently isolated via micro-segmentation, regardless of its physical location or underlying network topology. NSX micro-segmentation provides a foundational architectural shift to enable topology agnostic, distributed security services to applications in the evolving data center.

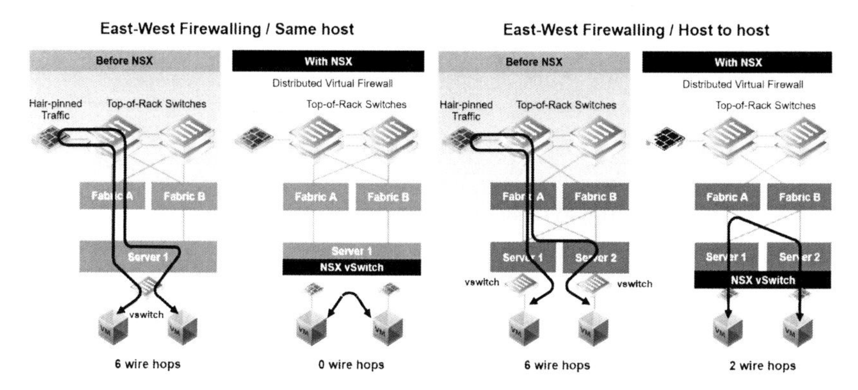

Figure 1.2 Perimeter-centric vs. NSX firewalling

Micro-segmentation Defined

Micro-segmentation decreases the level of risk and increases the security posture of the modern data center. Micro-segmentation utilizes the following capabilities to deliver its outcomes:

- **Distributed stateful firewalling:** Reducing the attack surface within the data center perimeter through distributed stateful firewalling. Using a distributed approach allows for a data plane that scales with the compute infrastructure, allowing protection and visibility on a per application basis. Statefulness allows Application Level Gateways (ALGs) to be applied with per-workload granularity.

- **Topology agnostic segmentation:** Providing application firewall protection regardless of the underlying network topology. Both L2 and L3 topologies are supported, agnostic of the network hardware vendor, with logical network overlays or underlying VLANs.

- **Centralized ubiquitous policy control of distributed services:** Controlling access through a centralized management plane; programmatically creating and provisioning security policy through a RESTful API or integrated cloud management platform (CMP).

- **Granular unit-level controls implemented by high-level policy objects:** Utilizing grouping mechanisms for object-based policy application with granular application-level controls independent of network constructs. NSX can use dynamic constructs including OS type, VM name, or specific static constructs (e.g., Active Directory groups, logical switches, VMs, port groups IP Sets). This enables a distinct security perimeter for each application without relying on VLANs.

- **Network based isolation:** Supporting logical network overlay-based isolation through network virtualization (i.e., VXLAN), or legacy VLAN constructs. Logical networks provide the additional benefits of being able to span racks or data centers while independent of the underlying network hardware, enabling centralized management of multi-data center security policy with up to 16 million overlay-based segments per fabric.

- **Policy-driven unit-level service insertion and traffic steering:** Enabling integration with third-party introspection solutions for both advanced networking (e.g., L7 firewall, IDS/IPS) and guest capabilities (e.g., agentless anti-virus).

Micro-segmentation and Cybersecurity Standards

NSX is the only micro-segmentation solution to have achieved the following industry standards:

- Common Criteria certification
- ICSA Labs certified firewall
- FIPS 140-2 certification (note - complete certification of all NSX cryptographic modules as of VMware NSX® for vSphere® 6.3)
- Validation in a published Micro-segmentation Cybersecurity Benchmark report by Coalfire, an independent cyber risk management advisor and assessor.
- Satisfies all NIST cybersecurity recommendations for protecting virtualized workloads (detailed below)

NIST Criteria

The National Institute of Standards and Technology (NIST) is a US federal technology agency working with industry to develop and apply technology, measurements, and standards. NIST works with standards bodies globally in driving forward the creation of international cybersecurity standards. NIST published Special Publication 800-125B, "Secure Virtual Network Configuration for Virtual Machine (VM) Protection" to provide recommendations for securing virtualized workloads. The micro-segmentation capabilities provided by NSX satisfy the security recommendations made by NIST for protecting virtual machine workloads.

Section 4.4 of NIST 800-125b makes four recommendations for protecting virtual machine workloads within modern data center architecture. These recommendations are as follows:

- **VM-FW-R1:** In virtualized environments with VMs running delay-sensitive applications, virtual firewalls should be deployed for traffic flow control instead of physical firewalls, because in the latter case, there is latency involved in routing the virtual network traffic outside the virtualized host and back into the virtual network.

- **VM-FW-R2:** In virtualized environments with VMs running I/O intensive applications, kernel-based virtual firewalls should be deployed instead of subnet-level virtual firewalls, since kernel-based virtual firewalls perform packet processing in the kernel of the hypervisor at native hardware speeds.

- **VM-FW-R3:** For both subnet-level and kernel-based virtual firewalls, it is preferable if the firewall is integrated with a virtualization management platform rather than being accessible only through a standalone console. The former will enable easier provisioning of uniform firewall rules to multiple firewall instances, thus reducing the chances of configuration errors.

- **VM-FW-R4:** For both subnet-level and kernel-based virtual firewalls, it is preferable that the firewall supports rules using higher-level components or abstractions (e.g., security group) in addition to the basic 5-tuple (source/destination IP address, source/destination ports, protocol).

NSX based micro-segmentation meets the NIST *VM-FW-R1, VM-FW-R2*, and *VM-FW-R3* recommendations. It provides the ability to utilize network virtualization based overlays for isolation and distributed kernel based firewalling for segmentation with API-driven centrally managed policy control.

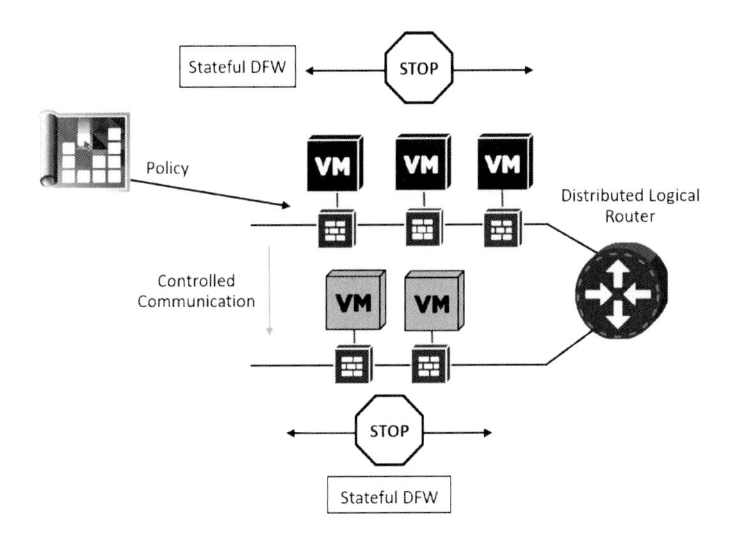

Figure 1.3 Distributed segmentation with network overlay isolation

Micro-segmentation through NSX also meets the NIST VM-FW-R4 recommendation for firewalling, utilizing higher-level components or abstractions (e.g., security groups) in addition to the basic 5-tuple (i.e., source/destination IP address, source/destination ports, protocol). NSX based micro-segmentation can be defined as granularly as a single component (e.g., application, IP address, virtual NIC, user) or as broadly as a region of multiple data centers. Controls can be implemented based on attributes of interest (e.g., identity, application).

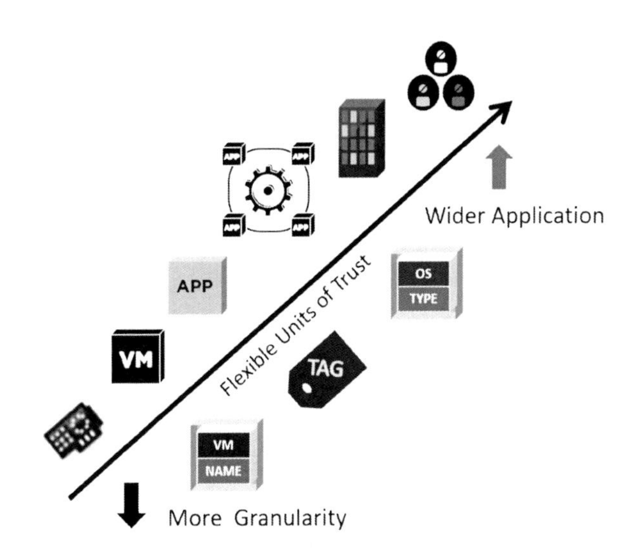

Figure 1.4 Flexible policy objects

Micro-segmentation with NSX as a Security Platform

Protection against advanced persistent threats that propagate via targeted users and application vulnerabilities requires more than L4 segmentation to maintain an adequate security posture. Securing chosen workloads against advanced threats requires application-level security controls such as application-level intrusion protection or advanced malware protection.

NSX based micro-segmentation goes beyond the recommendations noted in the NIST publication, enabling fine-grained application of service insertion (e.g., IPS services) to be applied to flows between assets that are part of a PCI zone. In a traditional network environment, traffic steering is an all-or-nothing proposition, requiring all traffic to be steered through additional devices. With micro-segmentation, advanced services are granularly applied where they are most effective; as close to the application as possible in a distributed manner while residing in separate trust zone outside the application's attack surface.

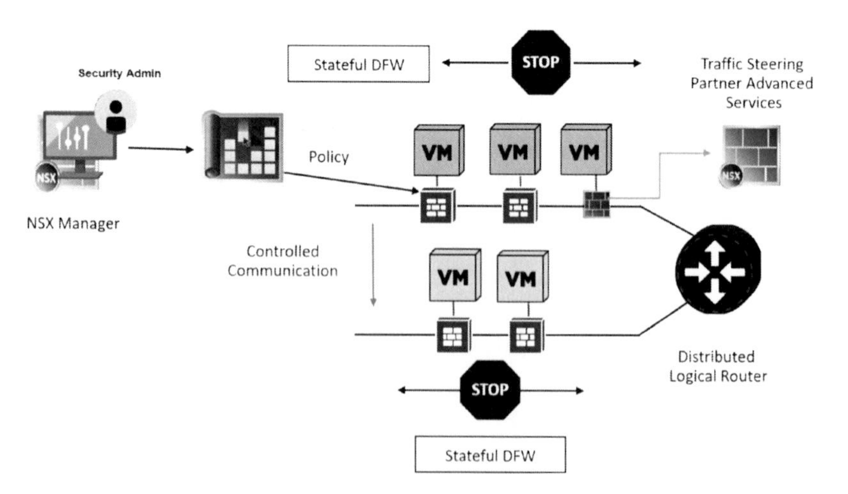

Figure 1.5 Distributed segmentation with network overlay isolation and service insertion

Securing Physical Workloads

While new workload provisioning is dominated by agile compute technologies such as virtualization and cloud, the security posture of physical workloads must still be maintained. NSX ensures the security of physical workloads; physical-to-virtual and virtual-to-physical communication can be enforced using Distributed Firewall rules at ingress or egress. For physical-to-physical communication, NSX can integrate automated security of static physical workloads through centralized policy control of those physical workloads via the NSX Edge Service Gateway or NetX integration with physical firewall appliances. This allows centralized policy management of both static physical and micro-segmented virtualized environments.

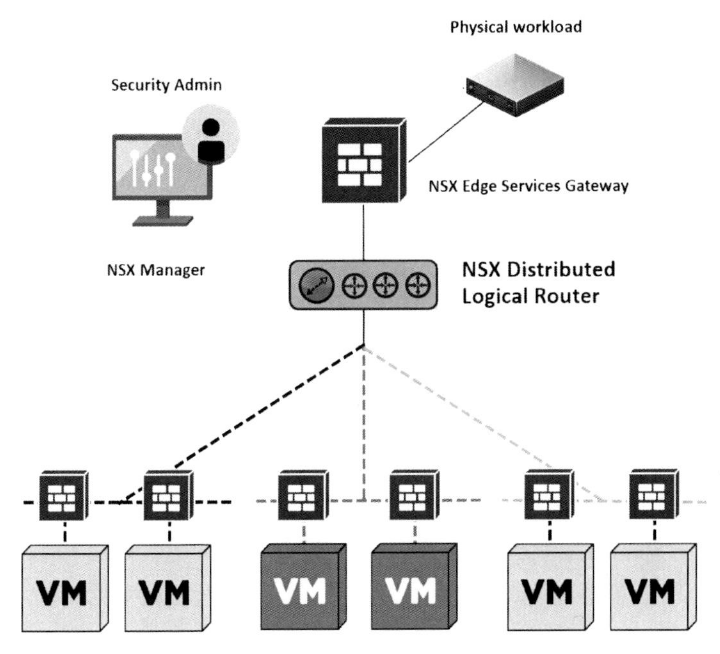

Figure 1.6 Secure virtual and physical workloads

Chapter 2

NSX Micro-segmentation Components

The VMware NSX platform includes two firewall components: a centralized firewall service offered by the NSX Edge Services Gateway (ESG) and the Distributed Firewall (DFW).

The NSX ESG enables centralized firewalling policy at an L3 boundary and provides layer 3 adjacencies from virtual to physical machines.

The DFW is enabled in the hypervisor kernel as a VIB package on all VMware vSphere® hosts part of a given NSX domain. It offers near line rate performance, virtualization, identity awareness, automated policy creation through Application Rule Manager (introduced in NSX for vSphere 6.3), advanced service insertion, and other network security features native to network virtualization. The DFW is applied to virtual machines on a per-vNIC basis.

Isolation

Isolation is the foundation of network security, whether for compliance, containment, or separation of development/test/production environments. Traditionally ACLs, firewall rules, and routing policies were used to establish and enforce isolation and multi-tenancy. With micro-segmentation, support for those properties is inherently provided.

Leveraging VXLAN technology, virtual networks (i.e., Logical Switches) are L2 segments isolated from any other virtual networks as well as from the underlying physical infrastructure by default, delivering the security principle of least privilege. Virtual networks are created in isolation and remain isolated unless explicitly connected. No physical subnets, VLANs, ACLs, or firewall rules are required to enable this isolation.

VLANs can still be utilized for L2 network isolation when implementing micro-segmentation with NSX, with application segmentation provided by the Distributed Firewall (see Figure 2.2). While using VLANs is not the most operationally efficient model of miro-segmentation, implementing application segmentation with only the DFW and keeping the existing VLAN segmentation is a common first step in implementing micro-segmentation in brownfield environments.

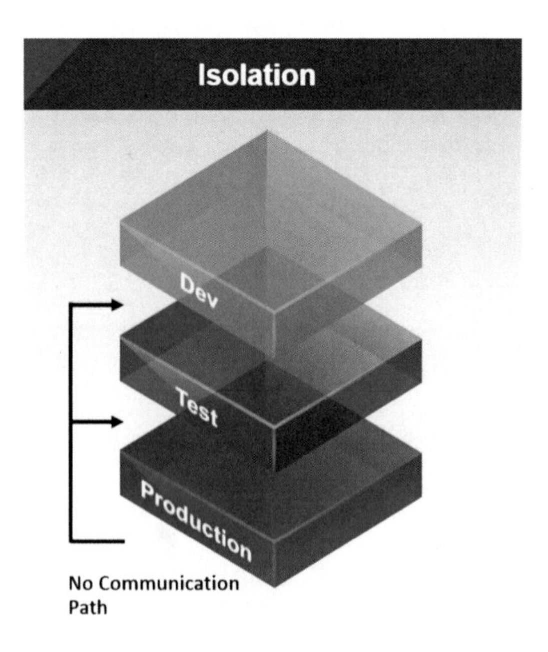

Figure 2.1 Network isolation

Segmentation

Segmentation, like isolation, is a core capability of NSX. A virtual network can support a multi-tier network environment. This allows for either multiple L2 segments with L3 isolation (Figure 2.2) or a single-tier network environment where workloads are all connected to a single L2 segment using Distributed Firewall rules (Figure 2.3).

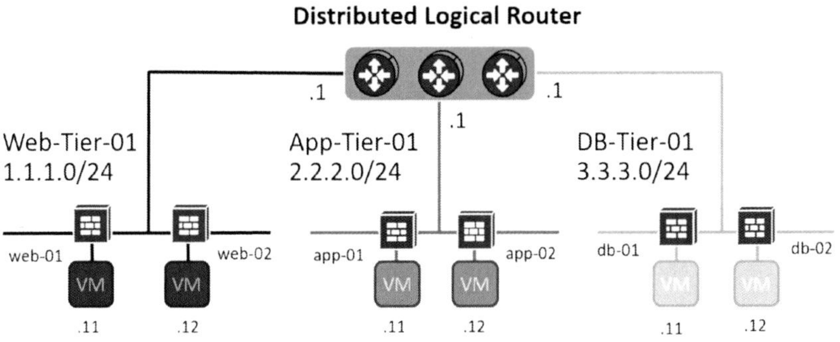

Figure 2.2 Multiple VXLAN L2 segments with L3 isolation

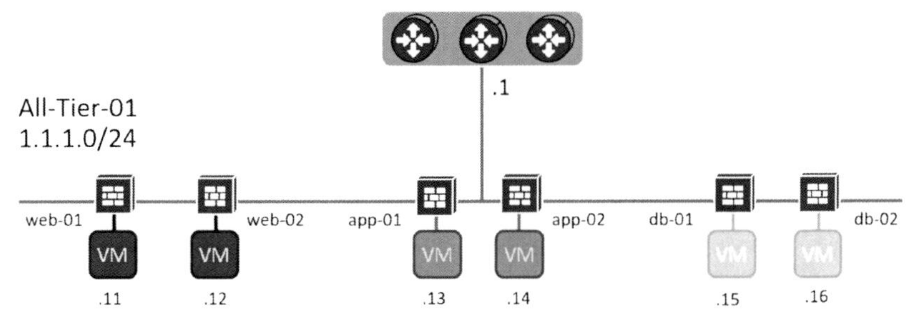

Figure 2.3 Single L2 segment with distributed firewall segmentation

Both scenarios achieve the same goal of micro-segmenting the virtual network to offer workload-to-workload traffic protection, also referred to as east-west protection.

Advanced Security Service Insertion, Chaining and Steering

Modern data center architectures decouple network and compute services from their traditional physical appliances. Previously, data center operation required traffic to be steered through appliances for services such as firewall, intrusion detection and prevention, and load balancing. As infrastructure services transition from physical appliances to software functions, it becomes possible to deploy these services with greater granularity by directly inserting them into a specific forwarding path. The combination of multiple functions in this manner is referred to as a service chain or service graph.

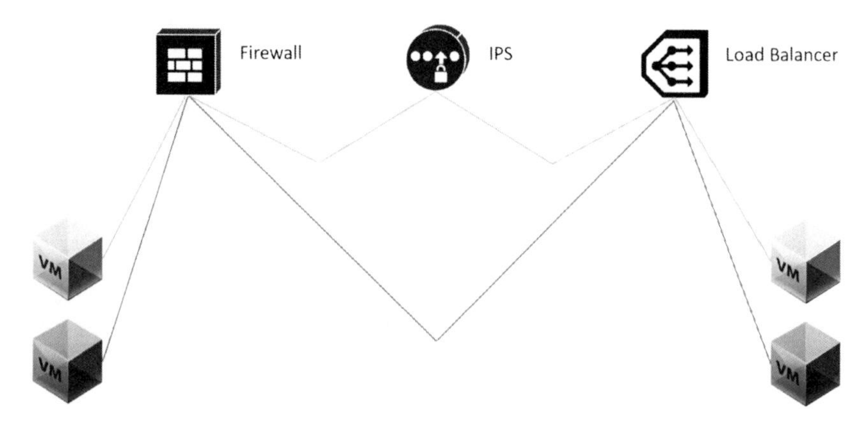

Figure 2.4 Two distinct service chains utilizing different functions

Once infrastructure services are defined and instantiated in software, they can be created, configured, inserted, and deleted dynamically between any two endpoints in the infrastructure. This allows the deployment and configuration of these services to be automated and orchestrated as part of a Software-Defined Data Center (SDDC).

The Role of Service Insertion in Micro-segmentation

Micro-segmentation allows for application-centric, topology-agnostic segmentation. Service insertion in this context permits granular security policies to be driven at the unit or application level rather than at the network or subnet level. This enables the creation and management of functional groupings of workloads and applications within the data center, regardless of the underlying physical network topology.

NSX provides L2-L4 stateful distributed firewalling features to deliver segmentation within virtual networks. In some environments, there is a requirement for more advanced network security capabilities. In these instances, organizations can leverage VMware NSX to distribute, enable, and enforce advanced network security services in a virtualized network environment. NSX distributes network services into the vNIC context to form a logical pipeline of services applied to virtual network traffic. Third- party network services can be inserted into this logical pipeline, allowing physical or virtual services to be equivalently consumed.

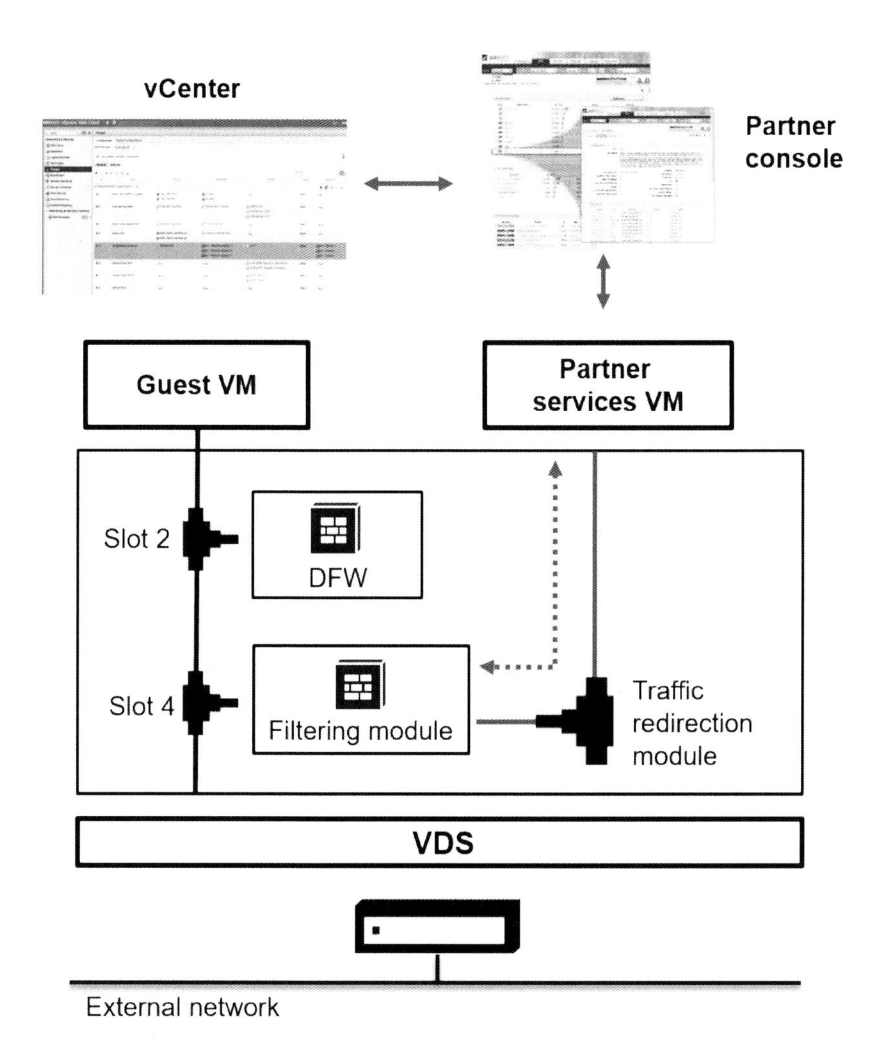

Figure 2.5 Service insertion, chaining, and steering

Between the guest VM and logical network (e.g., Logical Switch or DVS port-group VLAN-backed) there is a service space implemented into the vNIC context. Slot-ID materializes service connectivity to the VM. As depicted in Figure 2.5, slot 2 is allocated to the DFW and slot 4 to the specific third-party advanced security services. The remaining slots are available to plug in additional third-party services. Traffic exiting the VM is statefully inspected from L2-L4 at line rate, then redirected by policy to a partner solution for additional action (e.g., deeper L4-L7 inspection).

Every security team uses a unique combination of network security products to address specific environmental needs. Where network security teams are often challenged to coordinate network security services from multiple vendors, VMware's ecosystem of security solution providers already leverages the NSX platform. Another powerful benefit of the centralized NSX approach is its ability to build policies that leverage service insertion, chaining, and steering to drive service execution in the logical services pipeline. This functionality is based on the result of other services, making it possible to coordinate otherwise unrelated network security services from multiple vendors.

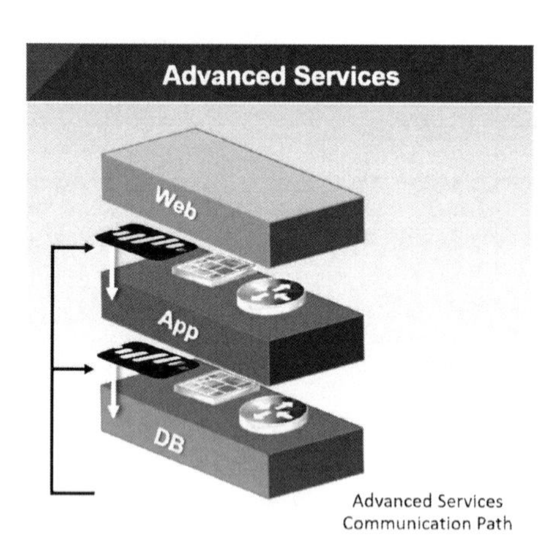

Figure 2.6 Network segmentation with advanced services provided by third-party vendor.

Integration with an advanced security services partner will leverage the VMware NSX platform to distribute the vendor's capability, making the advanced features locally available on each hypervisor. Network security policies, defined for applications workloads provisioned on or moved to that hypervisor, are inserted into the virtual network's logical pipeline. At runtime, the service insertion leverages the locally available advanced security service's feature set to deliver and enforce application, user, and context-based control policies at the workload's virtual interface.

Network and Guest Introspection

There are two families of infrastructure services that can be inserted into an existing topology: network services and guest services.

When deploying a network service, flows are dynamically steered through a series of software functions. For this reason, network services traffic may be referred to as data in motion. Network functions inspect and potentially act on the information stream based on its network attributes. These attributes could include the traffic source, destination, protocol, port information, or a combination of parameters. Typical examples of network services include firewall, IDS/IPS, and load balancing services.

Guest services act on the endpoints, or compute constructs, in the data center infrastructure. These functions are concerned with data at rest, primarily focusing on compute and storage attributes. Agentless anti-virus, event logging, data security, and file integrity monitoring are examples of guest services.

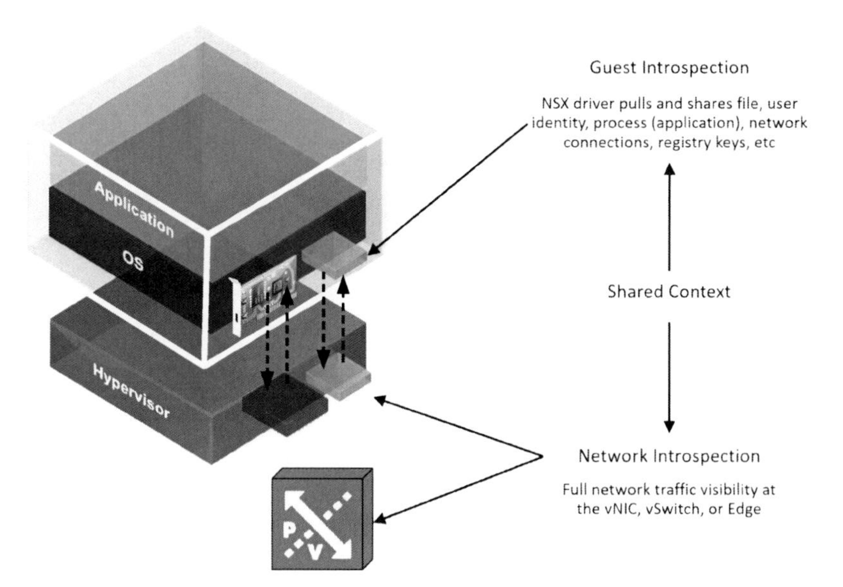

Figure 2.7 Network and guest introspection

NSX Service Insertion

As a hypervisor-integrated software platform, NSX provides a unique workload-centric insertion point of data center infrastructure services. Service insertion methodologies traditionally rely on network traffic steering to a set of software functions via the physical or logical network control plane. This approach requires an increasing amount of element management and control plane steering as the number of software services scales over time..

The NSX micro-segmentation based approach involves steering specified traffic via the NetX (Network Extensibility) framework through one or more Service Virtual Machines (SVMs). These SVMs do not receive network traffic through the typical network stack; they instead are passed traffic directly via a messaging channel in the hypervisor layer. Network traffic designated for redirection to a third-party service is defined in a policy driven manner utilizing an NSX feature called Service Composer. The traffic steering rules are based on defined security policies.

Packet are inspected and processed based on the applied rule set before reaching the network. In the example in Figure 2.8, this is a Virtual Distributed Switch port-group.

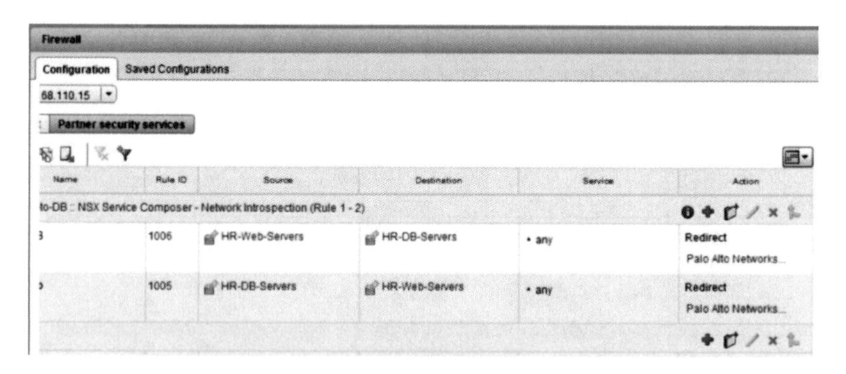

Figure 2.8 The NSX Distributed Firewall partner security services tab

This framework for traffic redirection is known as VMware Network Extensibility or NetX. The NetX program features a variety of technology partners across the application delivery, security, operations, and inter-domain feature sets, with additional partners constantly joining the ecosystem. NSX for vSphere 6.3 contains slots for up to eight different NetX services per VMware vCenter® cluster.

Security Benefits of Abstraction

Network security has traditionally been tied to network constructs. Security administrators had to have a deep understanding of network addressing, application ports, protocols, network hardware, workload location, and topology to create security policy. Network virtualization abstracts application workload communication from the physical network hardware and topology; this frees network security from physical constraints, enabling policy based on user, application, and business context.

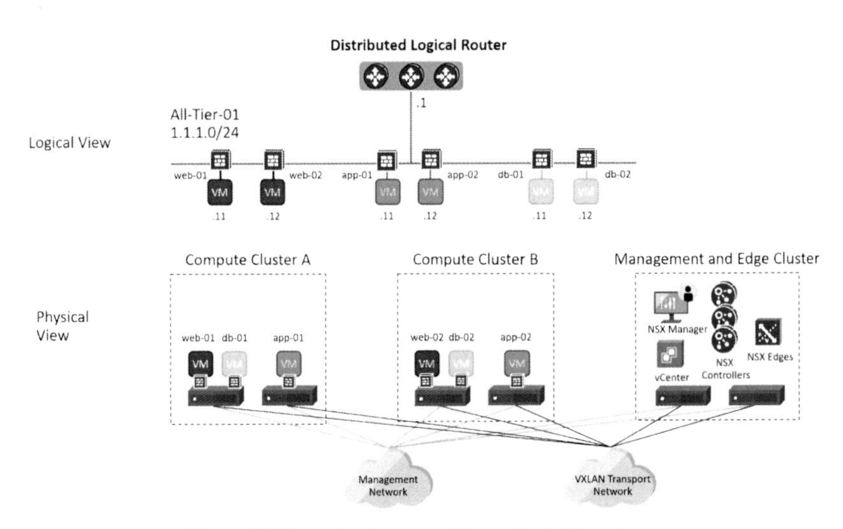

Figure 2.9 Security abstracted from physical topology

Service Composer

NSX micro-segmentation enables the deployment of security services independent of the underlying topology. Traditional (e.g., firewall) or advanced (e.g., agentless AV, L7 firewall, IPS, and traffic monitoring) services can be deployed independent of the underlying physical or logical networking topologies. This enables a significant shift in planning and deploying services in the data center. Services no longer need to be tied to networking topology. NSX provides a framework - Service Composer - to enable deployment of security services for the data center.

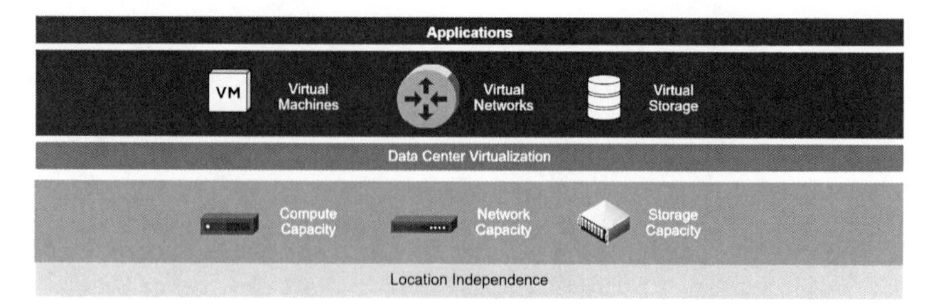

Figure 2.10 Software-defined data center

Service Composer consists of three broad parts:

- **Intelligent Grouping:** NSX decouples workloads from the underlying topology via creation of security groups.
- **Service Registration and Deployment**: NSX enables third-party vendor registration and technology deployment throughout the data center
- **Security Policies:** Security policies allow flexibility in applying specific security rules to distinct workloads. Policies include both governance of built-in NSX security services as well as redirection to registered third-party services (e.g., Palo Alto Networks, Check Point, Fortinet).

The simplicity of the Service Composer model is represented in the Figure 2.11.

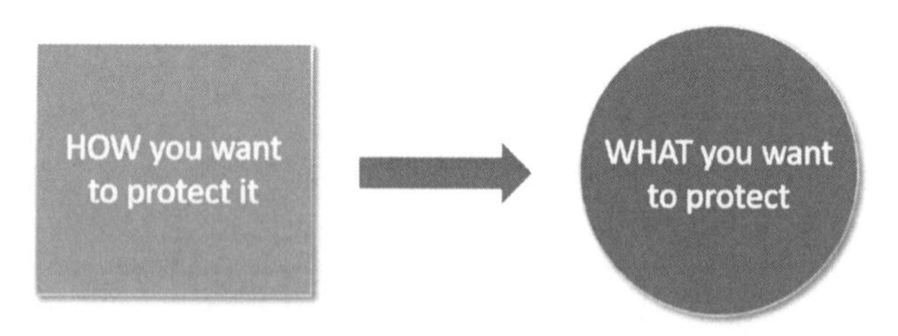

Figure 2.11 Decoupling of rules and policy

There are various advantages in decoupling the service and rule creation from the underlying topologies:

- **Distribution of Services:** The services layer of NSX allows distribution and embedding of services across the data center, enabling workload mobility without bottlenecks or hairpinning of traffic. In this model, granular traffic inspection is done for all workloads wherever they reside in the data center.

- **Policies are Workload-Centric:** Policies are natively workload-centric rather than requiring translation between multiple contexts - from workloads to virtual machines to basic networking topology/IP address constructs. Policies can be configured to define groups of database workloads that will be allowed specific operations without explicitly calling out networking centric language (e.g., IP subnets, MACs, ports).
- **Truly Agile and Adaptive Security Controls:** Workloads are freed from design constraints based on the underlying physical networking topologies. Logical networking topologies can be created at scale, on demand, and provisioned with security controls that are independent of these topologies. When workloads migrate, security controls and policies migrate with them. New workloads do not require the recreation of security polices; these policies are automatically applied. When workloads are removed, their security policies are removed with them
- **Service Chaining is Policy-based and Vendor Independent:** With NSX, service chaining is based on a policy across various security controls. This has evolved from manually hardwiring various security controls from multiple vendors in the underlying network. With NSX, policies can be created, modified, and deleted based on the individual requirements.

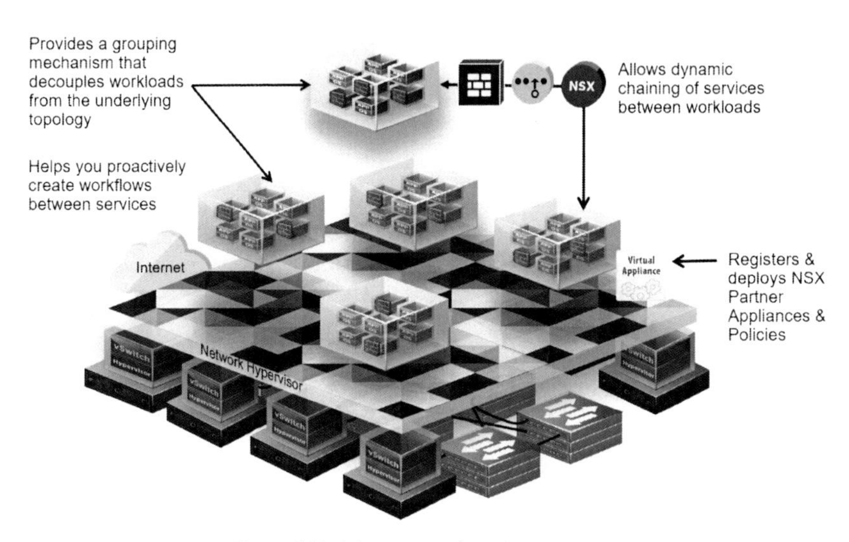

Figure 2.12 Advantages of service composer

Introduction to Intelligent Grouping

Intelligent grouping in NSX supports multiple creation displayed and a variety of customized grouping criteria; common approaches are shown in Figure 2.13.

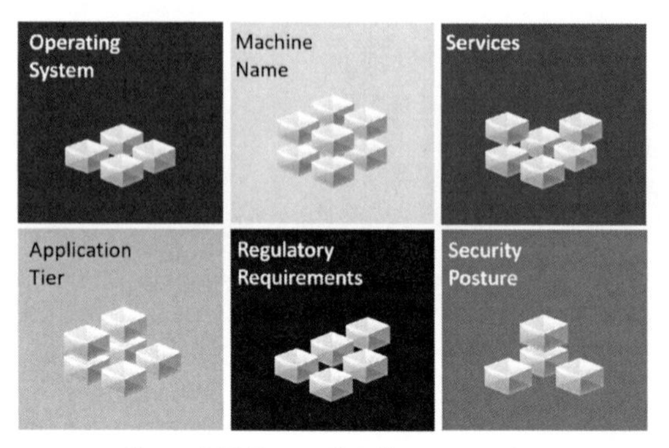

Figure 2.13 Types of intelligent grouping

Examples of grouping criteria include:

- **vCenter Objects:** VMs, Distributed Switches, Clusters, etc.
- **VM Properties:** vNICs, VM names, VM operating Systems, etc.
- **NSX Objects:** Logical Switches, Security Tags, Logical Routers, etc.

Grouping mechanisms can be either static or dynamic in nature, and a group may be any combination of objects. Grouping criteria can include any combination of vCenter objects, NSX Objects, VM Properties, or Identity Manager objects (e.g., AD Groups). A security group in NSX is based on all static and dynamic criteria along with static exclusion criteria defined by a user. Figure 2.14 details the security group construct and valid group objects.

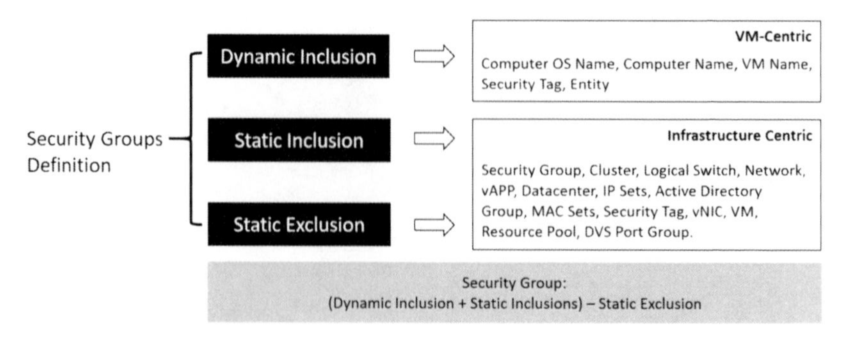

Figure 2.14 Scope of security of group attributes

Static grouping mechanisms are based on a collection of virtual machines that conform to set criteria. vCenter objects define data center components and NSX objects define core networking component. Combining different objects for the grouping criteria results in the creation of the AND expression in the NSX system.

Dynamic grouping mechanisms are more flexible; they allow expressions that define the virtual machines for a group. The core difference from static grouping mechanisms is the ability to define AND/OR as well as ANY/ALL criteria for the grouping.

Evaluation of VMs that are part of a group is also different. In a static grouping mechanism, the criteria instruct NSX Manager on which objects to look for in terms of change. In dynamic grouping, NSX Manager evaluates each change in the data center environment to determine which groups are affected.

As dynamic grouping has a greater impact on NSX Manager than static grouping, evaluation criteria should look at zones or applications and be mapped into sections to avoid the global evaluation of dynamic objects. When updates or evaluations are performed, only the limited set of objects belonging to a specific section are updated and propagated to specific hosts and vNICs. This ensures rule changes only require publishing of a section rather than the entire rule set. This method has a critical impact on performance of NSX Manager and the propagation time of policy to hosts. The sample sectionalized DFW rules set is shown in Figure 2.15.

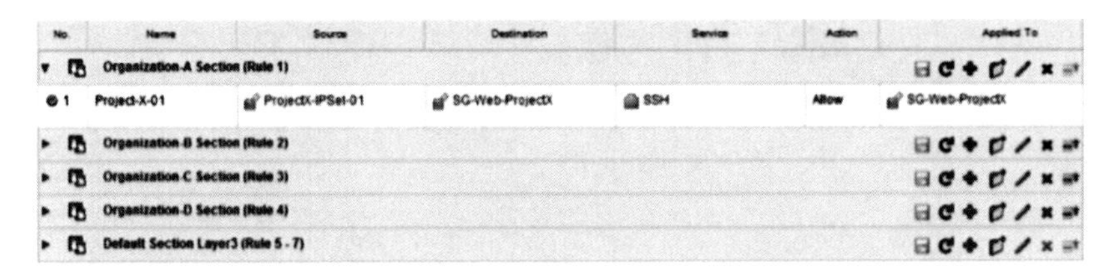

Figure 2.15 Creating actions for efficient rule processing and propagation

Intelligent Grouping – An Example of Efficiency

In the following example, virtual machines are added to a logical switch. In a traditional firewall scenario, IP addresses or subnets must be added or updated to provide adequate security to the workload. This implies that the addition of firewall rules is a prerequisite of provisioning a workload. Figure 2.16 diagrams this process.

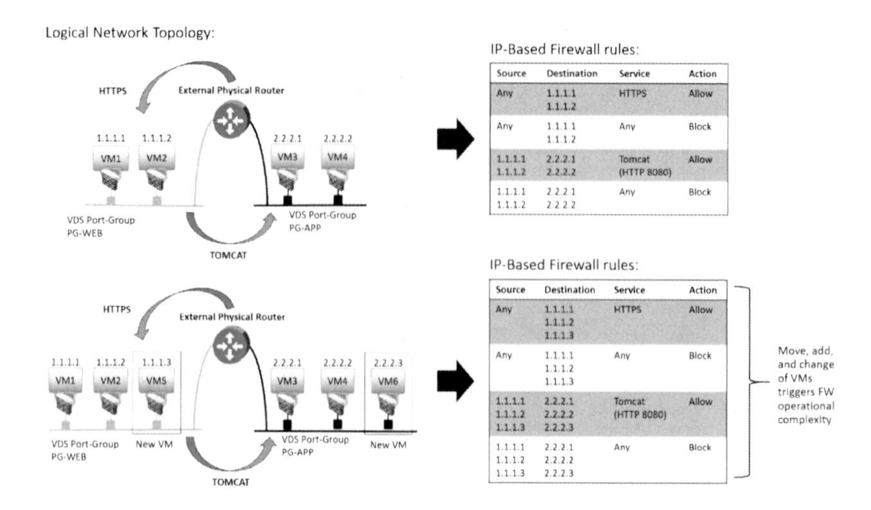

Figure 2.16 Traditional firewall rule management overhead

With NSX, a security or network administrator can define that any workload on this virtual machine will be assigned similar security controls. With intelligent grouping based on the NSX Logical Switch, a rule can be defined that does not change with every virtual machine provisioned. The security controls adapt to an expanding data center as shown in Figure 2.17.

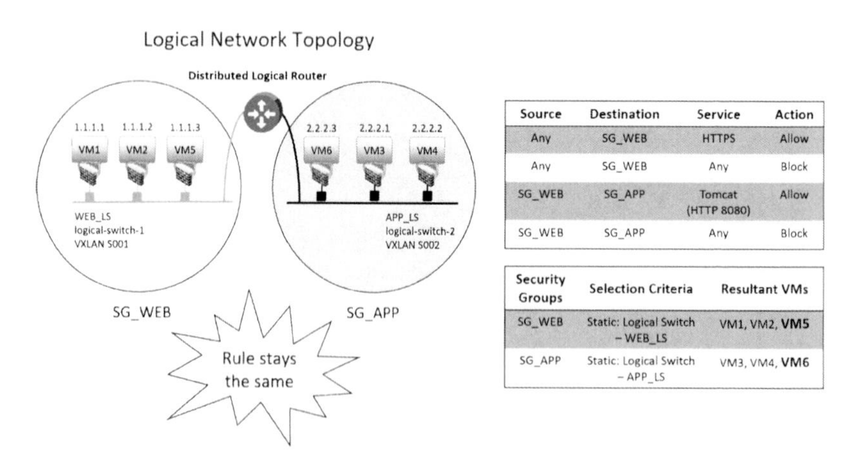

Figure 2.17 Adaptive security with NSX micro-segmentation

Security Tags

NSX provides security tags that can be applied to any virtual machine. This allows for classification of virtual machines in any desirable form.

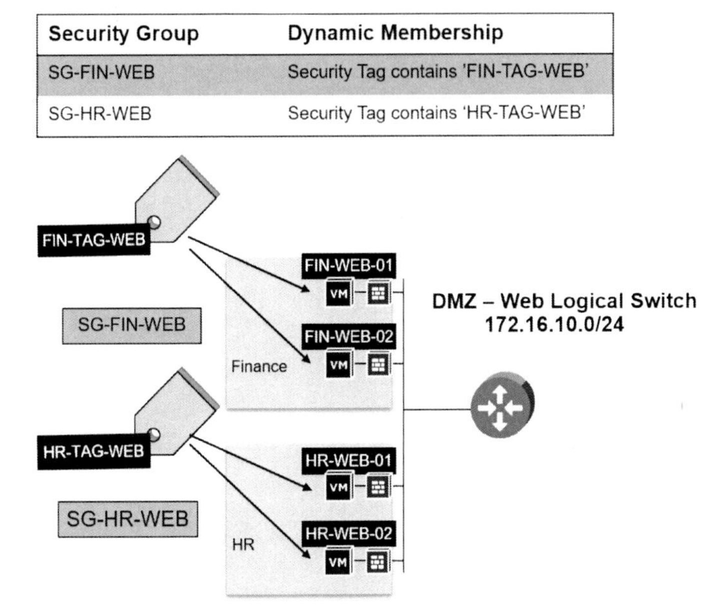

Figure 2.18 Tag used as base isolation method

Some of the most common forms of classification for using security tags are:

- Security state (e.g., vulnerability identified)
- Classification by department
- Data-type classification (e.g., PCI Data).
- Type of environment (e.g., production, devops)
- VM geo-location

Security tags are intended for providing more contextual information about the workload, allowing for better overall security. In addition to users creating and defining their own tags, third-party security vendors can use the same tags for advance workload actions. Examples of vendor tags include trigger on malware found, suspicious workload activity, and Common Vulnerabilities and Exposures (CVE) score compliance. This functionality allows for context sharing across different third-party vendors.

Chapter 3

Plan and Design for Micro-segmentation

A micro-segmented security architecture provides a foundation for enhanced security through granular isolation of and increased visibility into the data center. There are several factors to be considered when moving from legacy security constructs to policy driven micro-segmentation; these include the environments to be addressed with micro-segmentation along with models for security governance, policy, deployment, and consumption.

Operational Model

A security operational model is a representation of how an organization delivers secure services with a goal of decreasing risk. Security can never be addressed with technology alone; it requires a combination of people, process, and technology. NSX micro-segmentation improves operational models by decoupling security from physical network constructs, aligning it instead to application workloads. This allows security to be provisioned on-demand alongside the application.

While providing security context closer to the application improves the ability to provision security on-demand in an agile manner, the people and processes through which security policy is created and implemented via NSX must be considered. A common question is, "How do operational roles change within an organization when implementing micro-segmentation through NSX?" The degree of change will depend on the maturity of the current organizational structure.

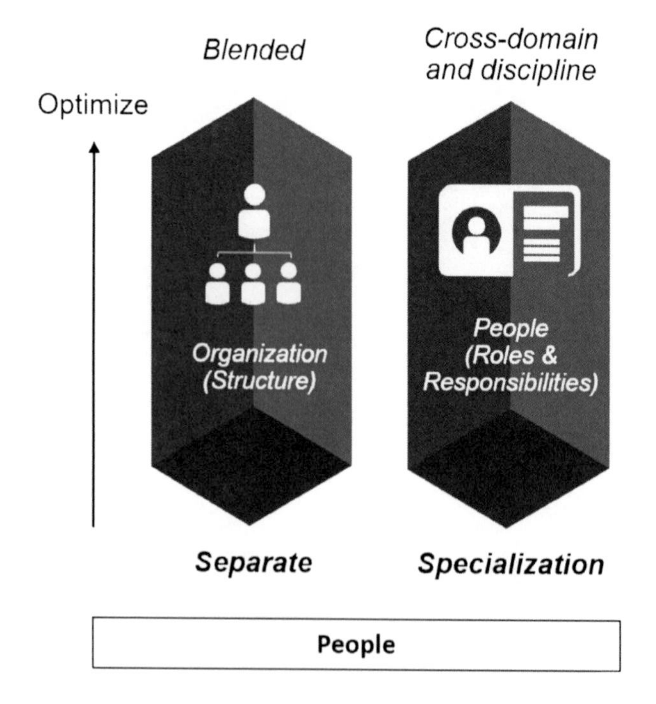

Figure 3.1 Optimize organizational structure

Organization-wide security mandates are dictated by the security team, but implementation of policy via NSX should be managed by a cross-functional team with representation from security, network, storage and compute. Implementation of micro-segmentation should start on a project basis, with policy applied to understood applications - either on a per-application or group-of-applications basis. Distinct challenges exist for new (i.e., greenfield) and existing (i.e., brownfield) environments. Tools such as NSX Application Rule Manager and VMware vRealize® Network Insight™ can provide the needed visibility into network flows to efficiently implement a micro-segmentation strategy; functionality and use of these tools is reviewed in Chapter 5.

Preparing Security Services for Data Center

NSX security services consist of built-in services (e.g., Distributed Firewall, Edge Firewall) and extensibility frameworks for enabling host and network based advanced services from third-party vendors.

Distributed Firewall rules are enabled on a cluster prepared for NSX by default, with default allow permit rules in place. This allows activation of the Distributed Firewall in each ESXi host, irrespective of whether it is a brownfield or greenfield environment. To enable DFW in a host, deploy the VIBs to that host (an automated, non-disruptive process accessible via NSX); this will enable a single DFW instance per vNIC on the VM.

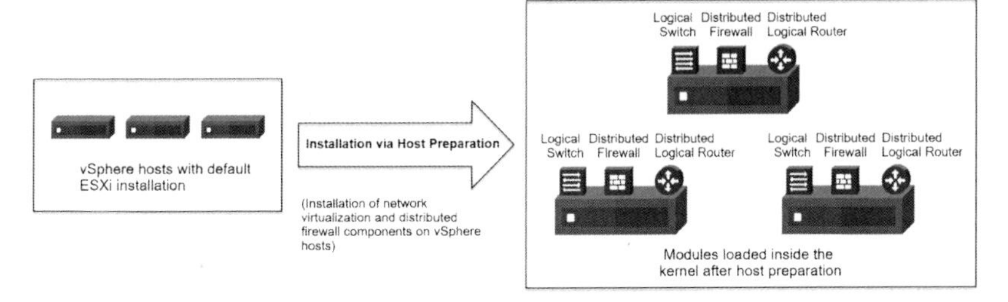

Figure 3.2 NSX VIBs installation to vSphere hosts

NSX Manager deploys VIBs to specified clusters. Once a cluster has been prepared, NSX Manager automatically prepares any new host subsequently added to the cluster. NSX components - NSX Manager, controllers, and Edge appliances - are automatically excluded from the DFW. vCenter is not automatically added to the DFW exclusion list, but it is recommended to do so manually.

Third-party security virtual appliances (SVAs) deployed via NSX are also automatically excluded from the DFW. Initial host preparation does not require a reboot of the host; upgrade scenarios are also reboot-free as of NSX for vSphere 6.3. Alerts and error messages are provided to offer visibility in case of a problem. The following best practices cover common scenarios in virtualized data centers.

Prepare Hosts in the Compute and Edge Clusters

vSphere data center designs generally involve 3 different types of clusters - management, compute and edge. VDI deployments may also have desktop clusters. Compute and desktop clusters must be prepared for the Distributed Firewall. Edge clusters may be prepared for the DFW if non-NSX Edge workloads will run in the edge cluster (common in VDI infrastructure scenarios).

The management cluster enables operational tools, management tools, and security management tools to communicate effectively with all guest VMs in the data center. Only enable the DFW on management clusters if east-west communication paths between these tools and the data centers are completely known. This ensures that management tools are not locked out of the data center environment. An example of a tool that can be used to enable a micro-segmented management cluster in an automated manner for VMware Horizon® virtual desktop infrastructure (VDI) deployments is the VMware Horizon Service Installer for NSX.

A separation of security policy between management components and workload is desired. It is recommended to exclude management components and security tools from the DFW policy to avoid lockout from east-west firewalling. Security tools with management servers must talk to their virtual appliances to provide signature updates, refresh rule sets, and initiate command and control tasks (e.g., scans). Asset management tools may need to deploy patches on virtual machines. In these scenarios, if the communication and data paths are known, then add DFW rules; however, while the port protocol needs are being examined and discovered it is recommended to exclude them from DFW to avoid disruption of essential controls.

Enable Specific IP Discovery Mechanisms for Guest VMs

NSX allows Distributed Firewall rules to be written in terms of object groupings that evaluate to virtual machines rather than IP addresses. NSX automatically converts the rules that contain virtual machines to actual IP addresses. To automatically update firewall rules with IP addresses, NSX relies on three different discovery mechanisms to provide IP addresses. Alternately, this can be updated using SpoofGuard.

Enable at least one of these IP discovery mechanisms:

- Deploy VMware Tools in guest VM.
- Enable DHCP snooping (NSX for vSphere 6.2.1 and above)
- Enable ARP snooping (NSX for vSphere 6.2.1 and above)
- Activate SpoofGuard with one of two different operation modes to update firewall rules with IP addresses:
- Manually authorize IP addresses for each VM.
- Enable trust on first use of IP address for a VM.

Note: All IP discovery mechanisms are available for both IPv4 and IPv6 addressing.

VMware Tools and ARP snooping enable learning of IP addresses of the virtual machine. This is effective where static IP addresses are configured manually on the guest VM. For DHCP-based IP address management, DHCP snooping is the correct choice. DHCP snooping also keeps track of lease timers and will properly expire IP addresses. DHCP snooping uses both DHCPv6 and Neighbor Discovery Snooping for IPv6 addresses.

When identity-based firewalling is used, VMware Tools deployment is recommended for guest introspection. VMware Tools detects not just the IP address of the VM but also the users logging in. Identity-based firewalling can also be achieved without VMware Tools by using Active Director log scraping.

VMware Tools and DHCP snooping are the recommended IP discovery mechanisms for guest VMs. Both mechanisms report changes in IP addresses as soon as they occur. They are also comprehensive in any environment, covering both dynamic and static addressing scenarios. If VMware Tools is used, care should be taken to ensure that it is always running. ARP snooping is also effective, but it should be used with SpoofGuard to avoid ARP poisoning.

The recommended approaches in selecting IP discovery are as follows:

- If VMware Tools is deployed, use VMware Tools and DHCP snooping for discovery of IP addresses.
- If VMware Tools is not deployed, use DHCP and ARP snooping for discovery of IP addresses.
- When using the Identity Firewall with guest-introspection, VMware Tools is required.

NSX provides a SpoofGuard feature to avoid spoofing of IP addresses. There are two SpoofGuard mechanisms: Trust on First Use (TOFU) and manual authorization.

TOFU will trust the first IP address reported to the NSX Manager via any of the described methods. For manual authorization, it will present the set of IP addresses discovered via any of the methods for approval by users. Users are still able to edit and add a different IP address.

Note that VMs initially obtain a link local IP address 169..*.*; ensure that is added to the list of trusted address via SpoofGuard policy.*

In a DHCP environment, TOFU is not recommended as the frequent IP address changes will cause operational challenges. Similarly, manual authorization of IP addresses will require automation in a large dynamic environment.

Policies are used to enable SpoofGuard for virtual machines. These policies are tied to logical switches or distributed port groups. For a single policy, all IP addresses approved must be unique. If overlapping IP addresses are desired, the virtual machines should be part of different SpoofGuard policies (i.e., part of different logical switches or distributed port-groups).

Note: SpoofGuard cannot currently be enabled on a NSX Cross vCenter Universal Logical Switch

Extensibility Framework for Advanced Security Services

NSX security fabric includes two extensible network frameworks (NetX and EPSec) that allow for the data center to use built-in and third-party vendor advanced security services. These extend NSX built-in segmentation and provide additional security services.

Extensible framework Guest Introspection enables host based security services like anti-virus, anti-malware, and Endpoint Monitoring. This framework is provided by a Guest Introspection Service appliance. As part of the Guest Introspection service appliance, NSX automatically deploys vShield Endpoint VIBs. The deployment considerations are the same as with the Distributed Firewall. Extensible framework network introspection enables network security services including L7 firewall and IDS/IPS. It is deployed as part of the Distributed Firewall and carries the same deployment considerations.

Deploying NSX Built-in Advanced Services

NSX provides two advanced detection services - Application Rule Manager and Endpoint Monitoring. Application Rule Manager simplifies the process of creating security groups and whitelisting firewall rules for existing applications. Endpoint Monitoring aids application owners in micro-segmentation planning by profiling applications and identifying the processes making network connections. These services are covered in greater detail in Chapter 4.

Deploying Third-Party Vendor Security Services

Deployment of third-party vendor security services like anti-malware, anti-virus, L7 firewall, or IPS/IDS requires considerations in addition to preparing the hosts for the DFW. NSX-integrated third-party security services contain two parts: the management console/server and a Security Virtual Appliance (SVA). NSX Manager requires connectivity with the third-party management server when creating policies. NSX Manager also requires connectivity to web servers where the OVF files for the security virtual appliances are stored.

Third-party vendor security managers should be deployed in the management cluster of the data center. NSX Manager deploys SVAs across the data center on every host. Deployment considerations for SVAs are the same as those for deploying the Distributed Firewall and preparing hosts across the data center. All vMotion considerations included for workloads protected by the DFW are also applicable for SVA deployments. The VM vMotion for NetX is handled by the DFW at the destination host so no TCP traffic drop will be seen. The SVA should never be vMotioned/Storage vMotioned.

Determining Policy Model

Policy models in a data center are essential to achieve optimized micro-segmentation strategies. They are required to enable optimum groupings and policies for micro-segmentation.

The first criteria in developing a policy model is alignment with the natural boundaries in the data center, such as application tiers, SLAs, isolation requirements, and zonal access restrictions (e.g., production/development, internal/external). Associating a top-level zone or boundary to a policy helps apply consistent and flexible control. Global changes for a zone can be applied via single policy, while within the zone there can be secondary policies with sub-group mappings to multiple sub-zones. An example production zone might be carved into sub-zones like PCI or HIPAA, or there may be multiple zones for VDI deployments based on user types. Various types of zones can be seen in Figure 3.3, where the connectivity and access space is segmented into DMZ, app, and database tier zones. There are also zones for each department as well as zone for shared services. Zoning creates relationships between various groups, providing basic segmentation and policy strategies.

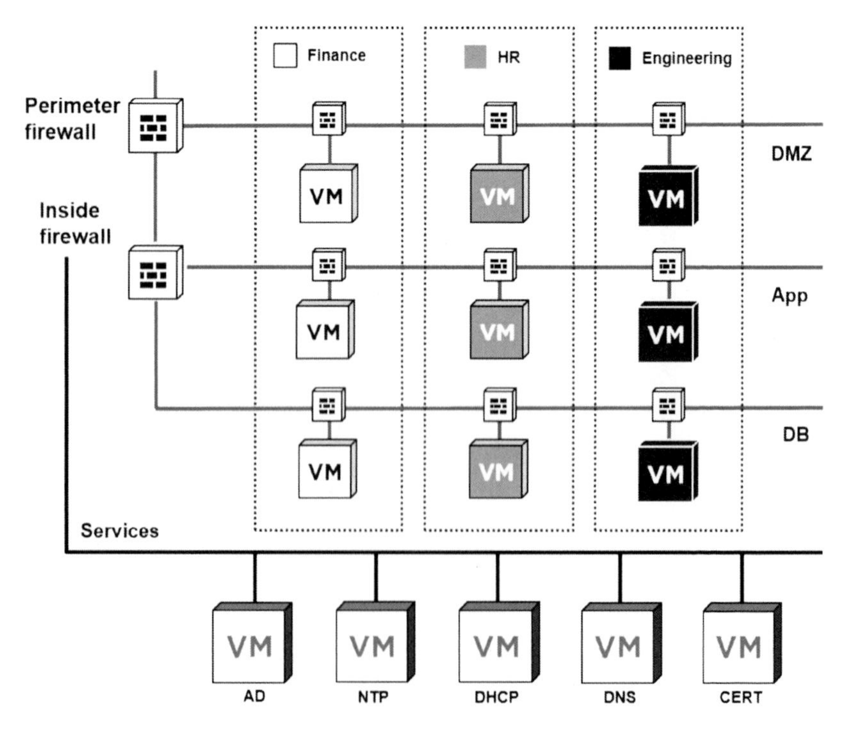

Figure 3.3 An example of various zones

A second criteria in developing a policy model is identifying reactions to security events and workflows. If a vulnerability is discovered, what are the mitigation strategies? Where is the source of the exposure - internal vs external? Is the exposure limited to a specific application or operating system version? Answers these questions to help shape a policy model.

Policy models should be flexible enough to address ever-changing deployment scenarios rather than simply be part of the initial setup. Common model concepts include intelligent grouping, tags, policy inheritance, universal vs. local policy, and hierarchy structure. A well-defined policy provides flexible and agile response capability for both steady state protection and instantaneous threat response.

A sample policy model is shown in Figure 3.4.

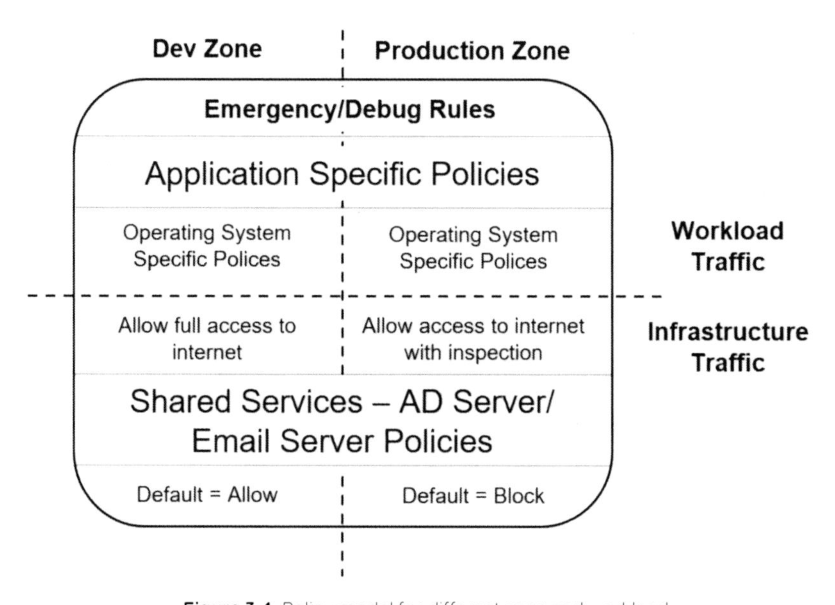

Figure 3.4 Policy model for different zone and workload

To effectively protect data center from attacks, a Zero Trust Model is recommended to implement a foundation of least privilege within the data center architecture. NSX micro-segmentation allows creation of this Zero Trust Model. A Zero Trust Model means only whitelisted applications are allowed access. The default rule will be a default deny policy. This model ensures that micro-segmentation rules will allow only specific traffic.

Implementing a Zero Trust Model in a brownfield environment requires knowledge of applications network flows; this can be derived by tools such as Application Rule Manager or vRealize Network Insight. If there is little insight into the east-west traffic of applications in a brownfield environment, it is recommended to monitor application traffic with one of the visibility tools covered in Chapter 5 before implementing a default deny rule (i.e., whitelist model).

Some organizations may be more comfortable with initially implementing NSX through a traditional blacklisting model. In this case, all traffic is allowed by default with administrators specifically restricting/blacklisting certain traffic. This model requires the administrator to block traffic that is viewed as unnecessary. It is simple to execute operationally, but difficult to ensure the best security posture. The best practice recommendation is to start with a default allow model and utilize the tools available to gain visibility and understanding of network traffic flows, enabling movement toward a default deny Zero Trust Model on a per application basis.

The quickest and the most efficient way to deploy policy is to create broad perimeters with a default set of restrictions, then provide additional protection with specifics sets of rules around zones. Once these broader zones and rules are created, each of the individual zones can be further analyzed to understand what specific rules are required.

Once the rules for the larger perimeters are completed, create a narrower perimeter around the application. Profile the application for its traffic pattern, then create more granular rules to control its traffic. Repeat this process for each application. This is an efficient strategy to implement micro-segmentation in a pragmatic manner.

Application profiling is the essential part of a micro-segmentation deployment. The aim of micro-segmentation is to enact a policy model based on either an open or Zero Trust Model, where workloads are protected from threats arising both internally and externally. A variety of tools and methods are available to expedite this process. Hardening guides are available for Many common applications have hardening guides available that specify various east-west and north-south communication paths. Collecting flow data using NetFlow can be helpful. Enabling DFW and logging all flows for analysis can provide a detailed profile of an application. Tools such as vRealize Network Insight, NSX Application Rule Manager, NSX Endpoint Monitoring, VMware vRealize® Log Insight™, VMware vRealize® Infrastructure Navigator™, and Application Dependency Planner can greatly simplify the application profiling process. Chapter 5 goes into further detail about some of these tools and processes.

Security Groups and Policies Design Considerations

This section examines design considerations for creating groups and policies along with nuances in deploying policy models to achieve higher efficacy. A further examination of grouping strategies based on a security group framework is discussed in Chapter 4.

Optimal Grouping and Policy

There are three factors central to creating an effective security policy: group nesting strategy, policy inheritance, and policy weights. Optimized policy and grouping requires a balance of all the three factors, helping better maintain the overall security posture of the data center. These factors are depicted in Figure 3.5

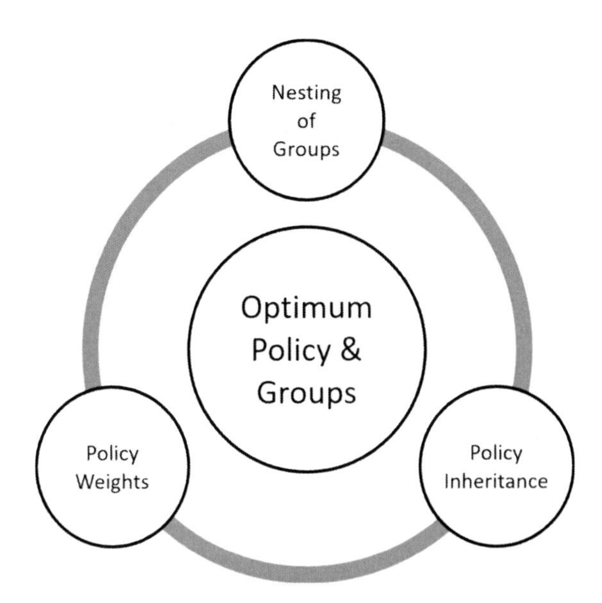

Figure 3.5 Factors affecting optimum policy & groups

Making Grouping Simple

To make grouping simple, policy models and weights must be well planned. A simple grouping strategy creates multiple groups per application or zone. VMs in this model ideally reside in mutually exclusive security groups (i.e., they are not nested). This approach requires the weight of each policy group to receive careful consideration to determine proper precedence. Additionally, policy rules will likely be complicated, sprawling similar rules across different policies.

Making Policy Weights Simple

The fewer the number of polices, the simpler the weight assignment. However, this strategy will create a lot of nested groups, and virtual machines will reside in multiple security groups. This increases complexity in both grouping and policy rule creation.

Making Policy Rules Simple

Creation of a minimum set of policy rules for protecting a given zone or application would ideally be operationally simple to understand. To make policy rules simple, the nesting of groups and policy weights must be well designed.

In the example case, web, app and database tiers require access to shared services (e.g., DNS, AD, SSO, NTP). If the policy is not nested, then all three tiers require distinct security groups, which then must be individually updated. Nesting of a group implies that a VM may to reside in multiple groups at the same time. This may cause rule evaluation for each VM to become overly cumbersome where the nesting is too deep; a best practice recommendation limits nesting to three to five levels.

With multiple policies, execution order should be efficient as policy weights translate to firewall precedence. Policy weights become crucial in determining the order in which the rules are applied. More nesting may make policy simpler to administer, while VM residence in multiple security groups will increase the complexity of policy weight. A well-balanced nesting depth and limited policy grouping is the optimal solution.

Group Creation Strategies

The most basic grouping strategy is creation of a security group around every application that is brought into the NSX environment. Each single tier, 2-tier, or 3-tier application should have its own security group; this will enable faster operationalization of micro-segmentation. When combined with a basic rule that says, "No application can talk to another except for shared essential services like DNS, AD, DHCP servers", this effectively enforces granular security inside perimeter. Once this basic micro-segmentation is achieved, writing rules per application will be desirable.

Creation of security groups gives more flexibility when the environment changes over time. Even for rules simply containing IP addresses, NSX provides a grouping object called an IP Set that can encapsulate IP addresses. This can then be used in security groups.

This approach has three major advantages:

1. Rules stay constant for a given policy model, even as the data center environment changes. The addition or deletion of workloads will affect only group membership, not the rules.

2. Publishing a change of group membership to the underlying hosts is more efficient than publishing a rule change. It is faster to distribute to the affected hosts and cheaper in terms of memory and CPU utilization.

3. As NSX adds more object grouping criteria, groups can be edited to better reflect the data center environment.

Use Grouping to Enhance Visibility

A virtual machine can be part of multiple groups. Groups can be used for data center visibility, resource categorization, and workload security rule application. A security group can contain all virtual machines that have Windows 2003 operating system. There might not be a security rule for all virtual machines with that operating system, but this grouping enhances the visibility of workloads in the data center. In this example, migration plans can be developed when the operating system is at its end of life or a specific vulnerability policy can be developed based on an announced or discovered security exposure.

Efficient Grouping Considerations

Calculation of groups adds a processing load to NSX Manager, with different grouping mechanisms adding different types of loads. Static groupings are more efficient than dynamic groupings in terms of calculation. At scale, grouping considerations should evaluate the frequency of group changes for a virtual machine. Large numbers of group changes that are frequently applied to a virtual machine means the grouping criteria is sub-optimal.

Using Nesting of Groups

Groups can be nested. A security group may contain multiple security groups or a combination of security groups and other grouping objects. A security rule applied to a parent security group is automatically applied to child security groups.

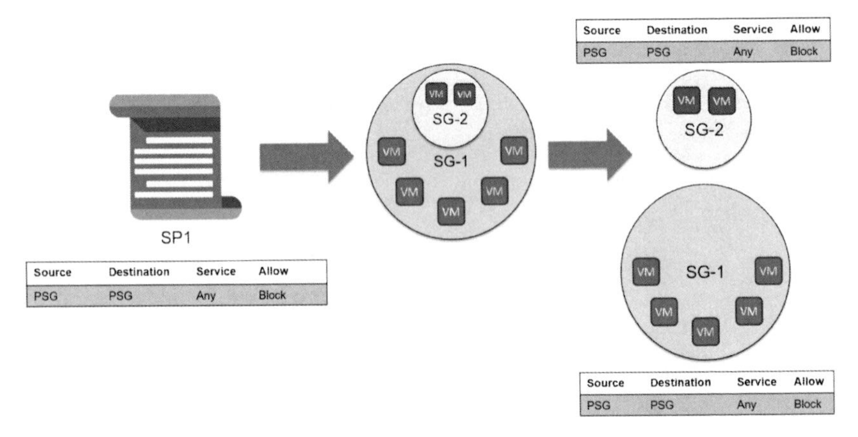

Figure 3.6 Policy nesting example

Using Policy Inheritance

Policy inheritances are generally costly operations for NSX Manager. While they use more processing power to compute the effective policy applied to the individual virtual machines, legitimate use cases exist. If a service provider/tenant model is used, the base policies developed by the service provider will form the guardrails for the tenants, while the child policies can be given to the tenants to create their own policies.

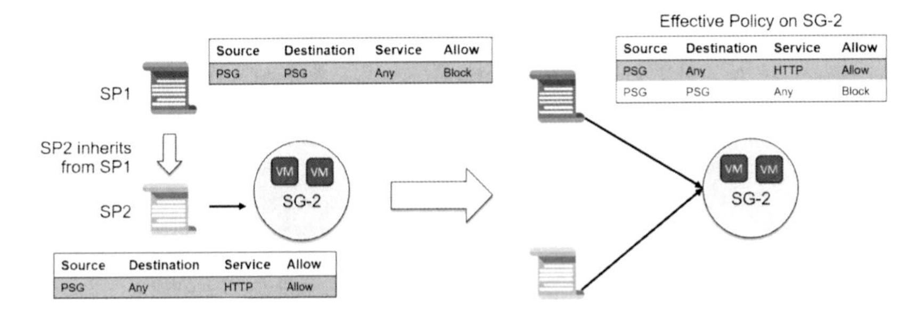

Figure 3.7 Policy inheritance example

Policy Creation Strategies

This section details the considerations behind policy creation strategies, helping determine which capability of the NSX platform should be exercised and how various grouping methodologies/policy strategies can be adopted for a given design.

Traditional vs. NSX Security Policy Approach

One significant question about deploying security with the NSX platform involves continued use of traditional security rules versus migration to the NSX security policy method.

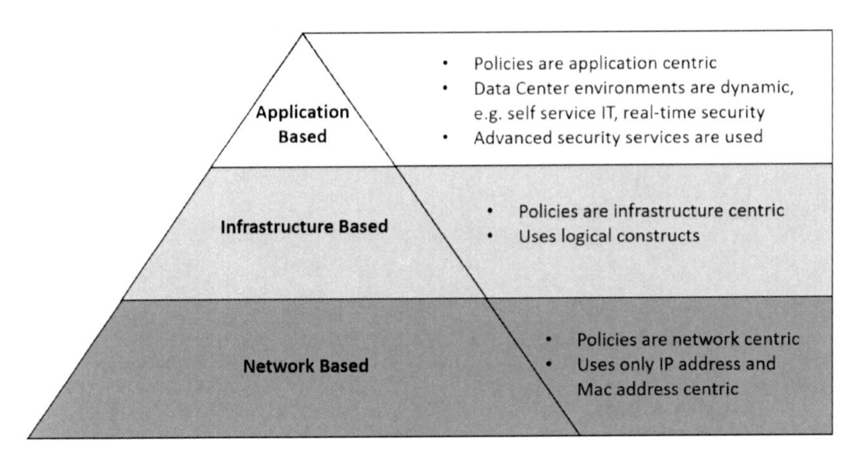

Figure 3.8 Context and hierarchy to policy creation

There are three general approaches for policy models showing in Figure 3.8; NSX provides flexibility in choosing a suitable model depending on environmental deployment and integration needs, each with specific advantages and disadvantages.

- **Network topology based policy models:** This is the traditional approach of grouping based on L2/L3 elements. Grouping can be based on MAC addresses, IP addresses, or a combination of both. Network models include both physical and logical topologies. Active-active, multi-site, and micro-segmentation extending to mobile applications through VMware Airwatch® integration will still utilize network policy based models.

  Advantages: This method of grouping works well for migrating existing rules from a different vendor's firewall environment.

  Disadvantages: The security team needs to be aware of the network topology to deploy network-based policies. There is a high probability of security rule sprawl, as grouping is not based on vCenter objects, NSX objects, or virtual machine attributes. The rules cannot be targeted to the workload; they are spread everywhere in the data center. This creates unnecessary configuration on every vNIC, leading to additional load on the data-plane.

- **Infrastructure based policy models:** In this approach, policies are based on SDDC infrastructure (e.g., vCenter clusters, logical switches, and distributed port groups). An example would mark cluster 1 to cluster 4 for PCI applications, with grouping based on cluster names and rules enforced across these groups. Another example would connect all VMs pertaining to an application to a specific Logical Switch.

  Advantages: Security rules are more comprehensive. The security team needs to work closely with the administrators that manage compute, networking, and storage in the data center. Workload abstraction using grouping is better than the network topology policy model.

  Disadvantages: The security team still must understand the logical and physical boundaries of the data center. Unlike the earlier policy model, it is also imperative to understand the compute and storage models. Workloads have physical restrictions on where they can reside and where they can move.

- **Application based policy model:** These policy models are developed for environments that are highly dynamic in nature. Multiple physical and logical network topologies may be created and removed on demand, potentially through automation or cloud consumption models to deploy workloads. Examples include self-service IT, service provider deployments, and enterprise private/public cloud.

  Advantages: Security posture and rules are created for the application, independent of the underlying compute, storage and network topologies. Applications are no longer tied down to network constructs or SDDC infrastructure, thus security policies can move with the application irrespective of network or infrastructure boundaries. Policies can be turned into templates and reused. Security rule maintenance is easier as the policies live for the application lifecycle and can be destroyed when the application is decommissioned. This practice helps in moving to self-service or hybrid cloud models.

  Disadvantages: The security team must to be aware of the interactions of the application that it is trying to secure. Application owners may need to document the security requirements of applications more extensively than before.

The following recommendations are provided to assist in developing optimal policy and grouping strategies.

- Nesting levels should be only 3 to 5 levels deep.
- Virtual machines should not be part of more than 10 mutually exclusive security groups.
- Virtual machines should not change security groups frequently.
- Policy inheritance should be kept at a maximum of 3 - 5 and serve only to create guardrails.
- Base policies in an inheritance should have mostly static criteria.
- Policy weights should be kept simple to streamline tracking and debugging of issues.

Deployment Models

There are multiple possibilities for workload location and migration; common use cases include:

- L3 to L7 policies across security zones
- Secure user environment using VDI/mobile infrastructure
- Tenancy model with service provider
- Protection between physical-to-virtual workloads
- Disaster recovery sites to back up data from primary data centers
- Remote Office/Branch Office (ROBO)

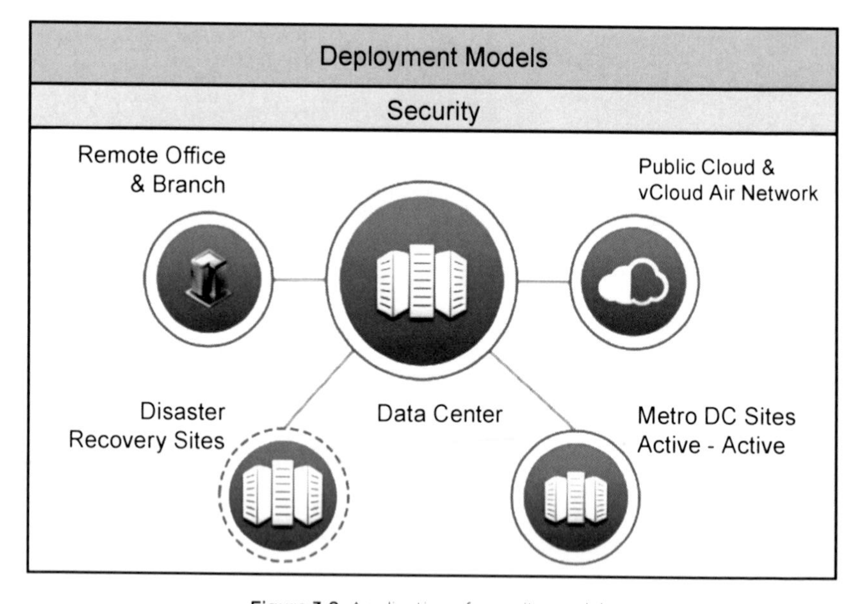

Figure 3.9 Application of security models

L3 to L7 Policies Across Zones

Data centers typically includes various zones for regulation (e.g., PCI, HIPAA), workloads (e.g., production, test, DMZ, application), users (e.g., developers, contractors, knowledge workers), or geographies (e.g., USA, EMEA, APJ).

Regulatory requirements or enhanced security edicts may not only require firewalling using L2-L4 rules, but additionally mandate L7 firewall rules and packet introspection technologies such as Intrusion Protection Systems (IPS).

When traffic moves across zones, it is recommended to use advanced security services from NSX partners to provide L7-based security technologies. Grouping strategies should create zones to provide logical demarcations. Policy strategies should include zone-specific policies that incorporate both built-in NSX security modules and third-party advanced services for enhanced security.

Secure User Environment using VDI

This model is pervasive in physical or virtual desktop service deployment. With NSX, a firewall can be adapted to individual users logging into the virtual machine. This feature is called the NSX Identity Firewall. NSX Identity Firewall allows intelligent grouping based on Active Directory user groups. NSX Identity Firewall also allows detection and determination of users logging into virtual machines, facilitating addition or removal of virtual machines from specific security groups. This capability allows secure micro-segments to be created based on user identity.

If Active Directory integration is not desirable or available in a data center, then security tags can be used to track users. For example, if Horizon View deploys and allocates a virtual desktop to a user, the virtual machine can be tagged with a security tag specifying the user details. Security grouping can be based on the tags that provide user-related information.

Mobile Infrastructure

NSX integrates with Airwatch Workspace One to allow end-to-end micro-segmentation from mobile application to data center application. In this model, the integration is achieved by mapping NSX security groups containing IP Sets to mobile applications in Airwatch.

Tenancy Model with Service Provider

In this model, the service provider creates guardrails to broadly constrain tenant operations; the tenant is then free to create its own rules within the sandbox provided. In a traditional approach, service providers can create sections for tenants. Tenants can then create rules in those sections and publish their own individual sections. A rule can further be tagged with keywords that enable logging to be provided only to the tenants for those rules.

In a policy approach, service providers can provide child policies to the tenants that inherit base policies. This ensures that the guardrails are covered in the base policy while tenants can update the child policies. If policy inheritance approach is not suitable in the design, then tenant policies can be of lower weights compared to service provider policies.

Protecting Traffic from Physical Workloads

Physical workloads are represented by their IP addresses in the NSX domain. Communication with physical workloads can be controlled in two different places in the NSX domain - through rules provisioned on the NSX Edge firewall or the Distributed Firewall.

Controlling Communication between VMs Across Multiple Data Centers

NSX allows universal firewall rules to be created and synchronized between vCenters. This is in addition to the local distributed rules allowed per vCenter.

Remote Office / Branch Office (ROBO) Model

An NSX domain can encompass hosts in both a central office along with multiple remote offices. This allows functionality such as micro-segmentation and service insertion to be extended beyond the data center. A single NSX Manager can be deployed at a central site to manage hosts prepared with NSX across many remote offices. Service Composer can be leveraged to build template policies and apply them to remote offices that require similar security. Furthermore, service insertion can be leveraged to deploy partner security services such as NGFW and IPS at branch locations; this is also managed centrally. Third-party software-based SD-WAN appliances can be used in conjunction with NSX at branch locations to provide per-application traffic steering and reduce the reliance on proprietary hardware and private circuits.

Cross vCenter Model

NSX Cross vCenter support enables a consistent security policy across vCenter boundaries and sites. This is accomplished by creating universal Distributed Firewall Rules. NSX provides centralized management on the primary NSX Manager for security rules across all vCenter domains/sites; users are no longer required to manually replicate security policies.

The NSX Rest API can also be leveraged so that a single call to the primary NSX Manager results in a consistent security policy across all sites for the application.

NSX Cross vCenter does not currently support utilizing Service Composer to create universal security groups or policies for cross vCenter traffic flows. Service Composer can be used to create local security groups to identify local workloads dynamically. This allows local security policies to be applied to workloads on universal networks. In this model the security configuration is created directly on both primary and secondary sites.

As of NSX for vSphere 6.3, security tag and VM name grouping criteria are synchronized across vCenters in an NSX cross vCenter deployment to support active-standby data center scenarios. In active-active models, IP sets and MAC sets are the only grouping criteria currently supported. For more information on cross vCenter design considerations, consult the NSX for vSphere Multi-Site Options and Cross VC NSX Design Guide.

Consumption Models

VMware NSX provides a RESTful API service via NSX Manager that can be consumed in several ways. The NSX REST API can be consumed directly via a tool/library such as cURL or a REST client like Postman, via multiple popular programming languages, and via orchestration or cloud management platforms. Popular programming languages such as Python, PowerShell, Perl, Go, and Java have REST client libraries can easily be utilized to consume the NSX REST API. Elaborate workflows and complete systems/portals can be created to provide custom automation, management, and monitoring capabilities, providing security as part of the application lifecycle.

Tools such as VMware vRealize® Orchestrator™ or programming language frameworks like Ansible can also be used to create advanced workflows for NSX. For additional details and examples see the Automation Leveraging NSX REST API Guide.

Cloud Management Platforms

Traditional ticket-based IT cannot match the increased agility that lines of business have come to expect from public cloud providers. Organizations want to free up time for business innovation rather than continue to spend resources on manual on-boarding and provisioning processes. While the self-service model is well understood for provisioning of workloads, configuration of appropriate networking and security remains a mostly manual process.

VMware vRealize® Automation™

NSX is integrated with vRealize Automation and supports additional integration with other cloud management platforms through the NSX RESTful API. With vRealize Automation, the provisioning of network and security services can be done in lockstep with application on-boarding. Security controls are deployed as part of the automated delivery of an application. The benefits of automation include:

- Faster application delivery through a standardized and repeatable process
- Greater reliability and consistency
- Reduce Opex and human errors by eliminating manual configuration tasks

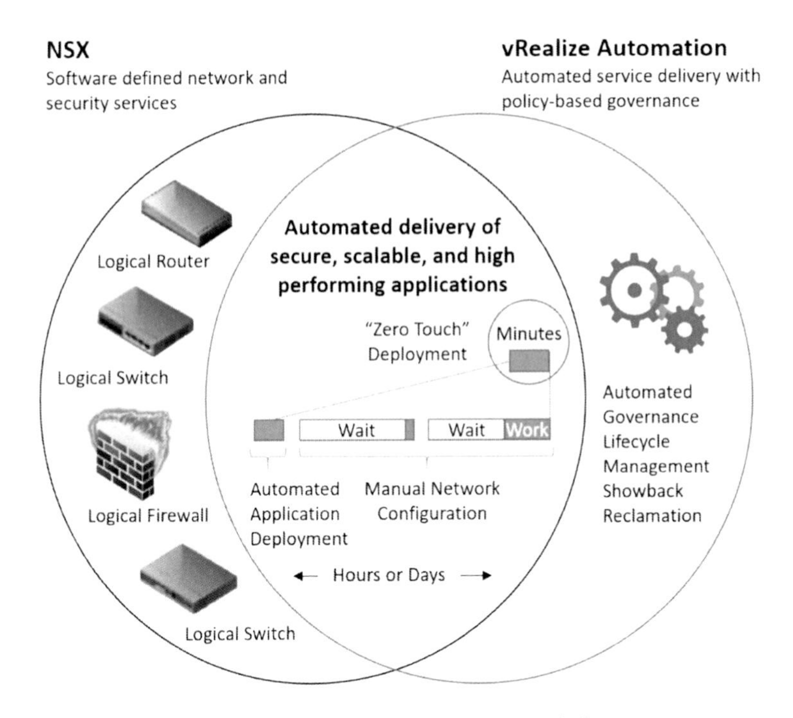

Figure 3.10 vRealize Automation and NSX

With vRealize Automation and NSX, an administrator can define application blueprints that specify NSX security policies for each application and application tier. These security policies include native Distributed Firewall rules as well as partner integration services such as L7 firewalling and agentless anti-virus.

Different options exist for automating application delivery micro-segmentation using vRealize Automation. One method uses security groups and policies representing application tiers that have been pre-configured in NSX. These pre-created policies should only allow inter-tier communication for specific services between each tier (e.g., allow MSSQL between the application and database tier). When creating a vRealize Automation blueprint, attach the application's workloads to its respective tiers. This approach ensures only controlled communication between application tiers is allowed when new applications are deployed from the blueprint.

Another option is to use the App Isolation feature inside of a vRealize Automation multi-machine blueprint. This is a simple checkbox will ensure security groups and policies are automatically created for every instance of the application that gets deployed, completely isolating this application from other applications or application instances.

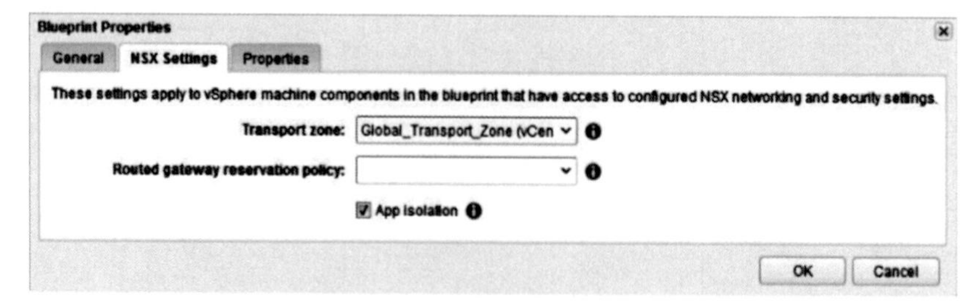

Figure 3.11 vRealize Automation app isolation checkbox

When creating a blueprint, an administrator can choose to use on-demand security groups and rules for each application instance. In this approach, define security policies in NSX but do not assign them to any security groups. When a multi-machine blueprint is defined in vRealize Automation, attach on-demand security groups to the application tiers and select the relevant security policy. Every time an application is deployed from this blueprint, a unique security group will be created. This isolates each application instance while at the same time micro-segmenting each application instance by use of the pre-configured policy.

VMware vCloud Director®

vCloud Director is VMware's multi-tenant infrastructure-as-a-service cloud management portal for service providers. As of vCloud Director 8.20, NSX security capabilities such as dynamic routing and Distributed Firewall are integrated and exposed through the self-service vCloud API. This enables service provider to allow tenants to manage and configure NSX Distributed Firewall based micro-segmentation on a per-tenant basis.

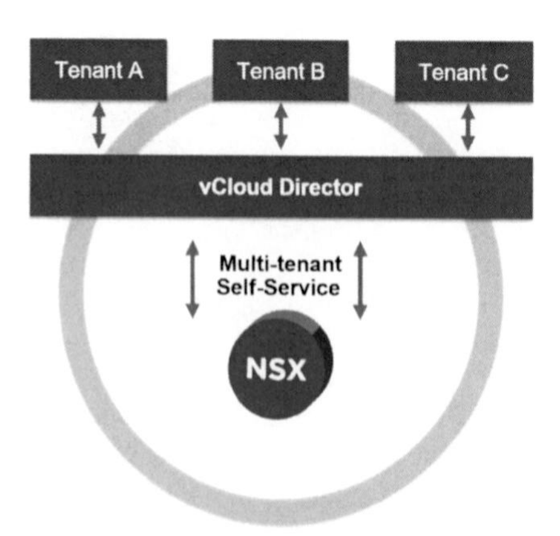

Figure 3.12 Per-tenant NSX capabilities exposed through the self-service vCloud API or vCloud Director portal

OpenStack

NSX is integrated with OpenStack, with the attributes of micro-segmentation fully available to the OpenStack admins and tenants. This includes leveraging the stateful Distributed Firewall and all service insertion capabilities with OpenStack as the cloud management platform.

Redefining the DMZ

Using NSX micro-segmentation to enable a DMZ Anywhere demonstrates how a proven security concept (e.g., traditional DMZ) can be brought into the modern SDDC, delivering service virtualization, cloud scalability, Zero Trust security, ephemeral servers, and applications-on-demand.

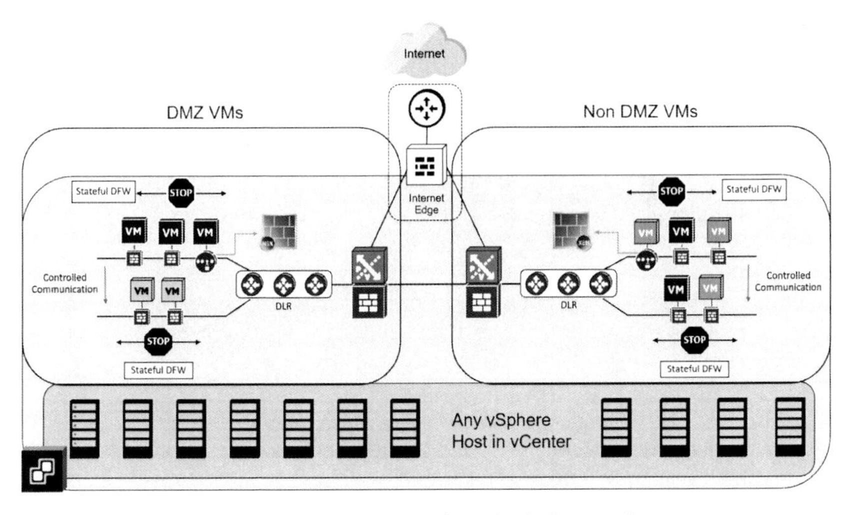

Figure 3.13 DMZ abstracted from physical constraints

Physical Security in a Virtual World

A mixed workload environment is one utilizing multiple application deployment models, including applications deployed on both virtual machines and legacy physical servers. NSX micro-segmentation satisfies the necessary security requirements for mixed workload environments through:

- Defining security requirements based on application deployment model or environment type
- Understanding methods of protection in modern data centers
- Designing NSX micro-segmentation for both physical and virtual workloads
- Integrating with ecosystem security and network controls functions

Security Requirements Differ in Heterogeneous Environments

Due to the evolving threat landscape and growing sophistication of cyber-attacks and threat actors, a single static policy or blanket approach to securing modern data centers is no longer adequate. These types of policies are difficult to manage and take a narrowly-focused approach to what needs to be a much broader solution. Today's private cloud environments are comprised of a variety of workloads and deployment models: different applications, distinct operating systems, and varied system platforms of physical servers and virtual machines. There are situations where platforms cannot be changed or optimized, such as legacy workloads tied to specific hardware. Inherent differences in architectures and deployment methods make having a single security policy and method of enforcement infeasible. The characteristics of the platform are important to determining which security solutions and policy attributes are applicable to each type of workload. There are distinct differences between legacy physical workloads versus those running on virtual machines within a software-defined data center; the physical workloads are static in nature while virtual/cloud workloads can be extremely dynamic.

Table 3.1 Physical vs. Virtual Workload Characteristics

Physical Workload	Virtual Workload
Static in nature with minimal network or system changes	Dynamic in both configuration and location
May have stricter security requirements around physical segmentation/ separation	Can easily scale up or down based on application and load requirements
Inability to move to a more modern platform due to legacy application dependences	Allows for security controls to be placed closer to the workloads with in-kernel hypervisor solutions

Security Control Placement

Deployment of platform-native security controls and integration with other ecosystem solutions allows for optimal placement of enforcement points. To achieve the highest level of control for data center traffic, it is essential security controls are placed as close to the workload as possible. Reducing the number of network hops or different transit devices a potentially malicious packet traverses helps minimize the exposure of other data center objects. By cutting the packet or traffic off closest to its origin, potential malicious activity can be mitigated or eliminated.

Protecting Mixed Workloads – Physical and Virtual

As a flexible and comprehensive security platform is needed to meet the requirements of modern data centers and applications, a baseline level of acceptable security must be defined to verify that the solution meets the needs of heterogeneous mixed workload environments. Key capabilities of an acceptable security solution include:

- A comprehensive set of policies and services for all workloads, both physical and virtual
- Stateful and application/workload-aware protection
- Ability to be fully automated
- Visibility into users, data center objects, and network traffic
- Granular enforcement at the level of units of compute or individual users
- Open APIs that allow consumption of security capabilities by higher level orchestration platforms

Providing Complete Micro-segmentation with NSX

To address essential security requirements, a solution must have deep integration at necessary protection points, complete visibility and control of traffic flows, and a flexible automated management and policy layer. NSX provides this level of security for all workloads - both physical and virtual - within a data center. As traffic flows continue to shift from traditional north/south network flows to east-west server communication, it is essential to enforce security at these intra-data center boundaries.

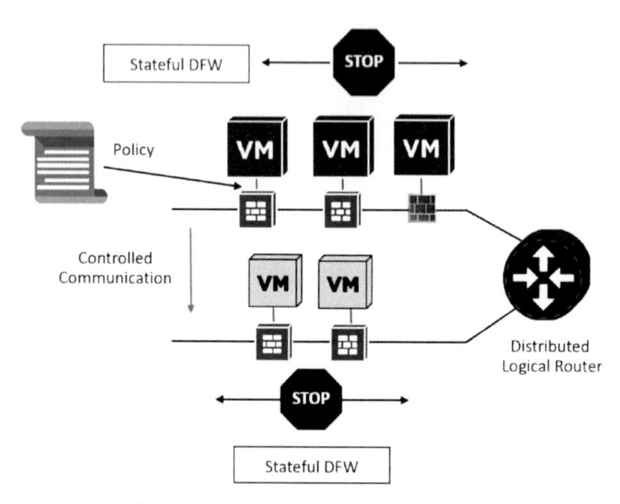

Figure 3.14 Distributed segmentation with network isolation

Using NSX Distributed Firewall, east-west communication between physical and virtual resources can be controlled without VLAN or network changes. This simplifies deployment into existing environments while offering uniform protection and inspection on critical data center traffic. In this model, an Edge Services Gateway provides north/south control for upstream physical server communication. To monitor and enforce east-west traffic between physical and virtual workloads, NSX places security controls at the hypervisor level. To achieve the tightest control over traffic flows, it is essential to place the security closest to the workload while still residing in a separate and secure trust zone based on the hypervisor. With this implementation, NSX places ingress enforcement at the receiving hypervisor and egress enforcement at the source hypervisor, providing end-to-end protection for layer 2 traffic flows. The model is flexible, supporting mixed enforcement methods for layer 3 deployments. This includes utilizing enforcement at the hypervisor level before the traffic is sent or leveraging an Edge Services Gateway to provide enforcement at the routing level.

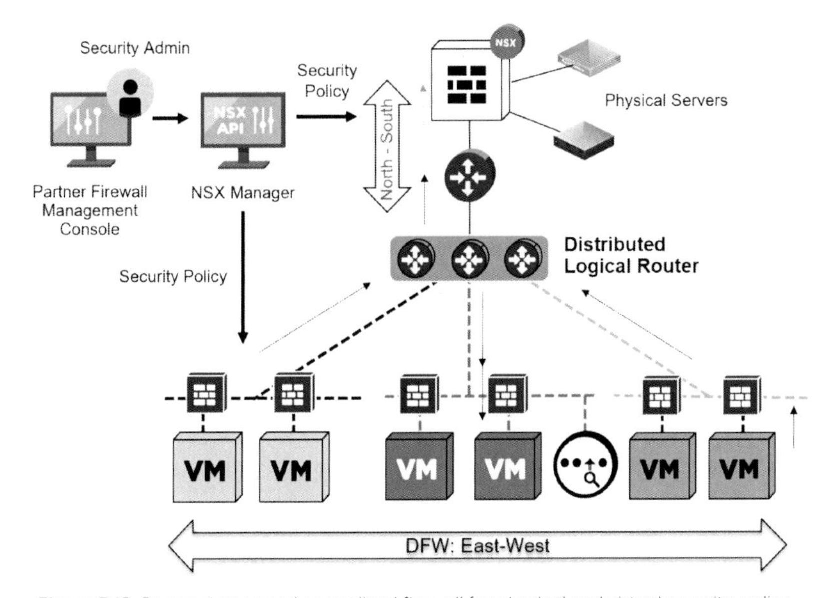

Figure 3.15 Partner integrated centralized firewall for physical and virtual security policy control

Policy definition and object grouping capabilities are important aspects of an operationalized micro-segmentation deployment. Without these abilities, a solution that granularly secures assets rapidly becomes unmanageable. NSX provides the ability to define security groups and IP Sets for easy grouping of objects included in a policy. The definition of these groups can be done manually or dynamically, with dynamic discovery criteria allowing administrators to quickly identify and group workloads based on characteristics such as operating system, application type, name, or network attributes.

Effective automation is also required to properly maintain rule bases and policies as environmental components (e.g., workloads, users, systems) constantly change. NSX uses dynamic discovery and grouping to ensure the most restrictive firewall ruleset is in place. When a new server - physical or virtual - is created, NSX automatically adds this system to the appropriate group and applies the necessary policies based on predefined attributes. As systems go offline or are decommissioned, stale rules are automatically cleaned up. This eliminates the "Swiss cheese" problem (i.e., holes in defense due to inadequate rules maintenance) seen with manually managed firewalls.

As a true security platform, NSX provides open APIs and built-in integration with third-party enterprise security solutions including policy and firewall management, access and change control tracking, advanced next-gen firewalling and threat protection, and anti-virus and IPS/IDS. This integration allows for the protection, management, and grouping components of NSX to be extended to pre-existing solutions within data centers.

Native Integration with Physical Networks and Hardware

Integration with existing physical network and security devices is important for solutions deployed in mixed workload environments. Native integration with top-of-rack devices and core network hardware is beneficial to achieve high levels of performance, port density, and policy management across large/service provider scale data center environments. This integration provides low-latency, data center-wide scaling of VXLAN termination natively on physical hardware. Whether for performance, security appliance capability, or tag/information sharing purposes, support for these devices ensures interoperability and allows access to the unique features that each solution offers.

NSX natively provides software layer 2 bridging to physical environments. Additionally, NSX provides layer 2 gateway integration through technology partners. This integration supports direct VXLAN termination on top-of-rack switches via hardware VTEPs, extending both the security and networking capabilities natively into these networks. When working across heterogeneous devices, NSX continues to provide centralized management for policies across supported sources. It also offers the ability to orchestrate and automate the ongoing creation and management of these objects.

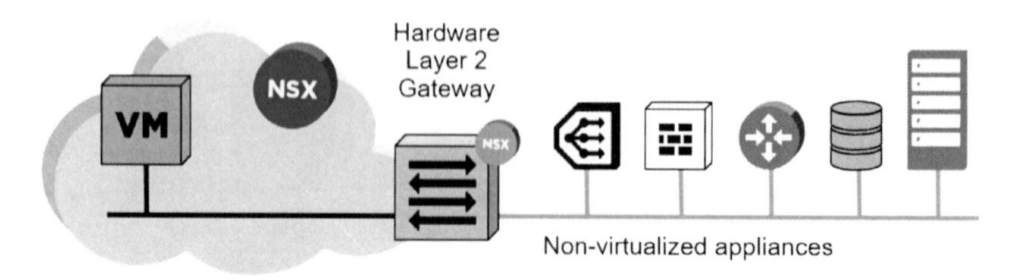

Figure 3.16 Software or hardware L2 extension

Deployment Example

NSX provides significant flexibility through multiple deployment models. In this example, a virtual ESG provides routing and firewalling functions for virtual and physical workloads. This provides tight access control via micro-segmentation at the hypervisor level while also protecting east-west traffic passing between virtual and physical machines. North-south traffic from the physical environment can also flow through this virtual endpoint for a single point of enforcement. If an existing physical firewall or security device is already in use, NSX and the ESG can be used to protect east-west communication while continuing to leverage the physical appliance for north/south traffic.

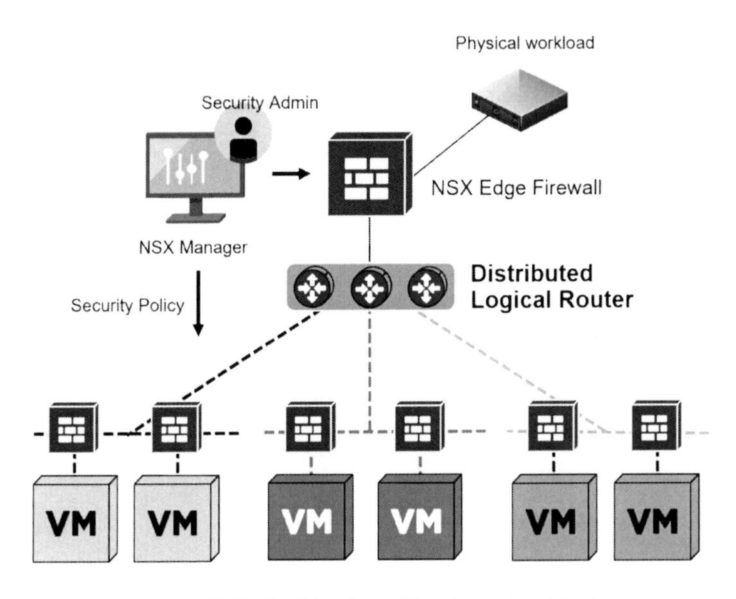

Figure 3.17 NSX Edge firewall for physical workloads

For a complete set of sizing, deployment, and feature guidelines, please see the NSX for vSphere Design Guide.

Modern data centers will continue to have mixed workload types and deployment models. This reality highlights the need for a security solution such as NSX that supports flexible policies and multiple enforcement methods based on the characteristics of the objects that needs to be protected. The key to effectively protecting physical and virtual workloads is to maintain east-west traffic control in a scalable and manageable manner. Support for integration with ecosystem solutions further enhances security context within the chain of protection, facilitating additional levels of security coverage.

Consistent Visibility and Security Across Physical and Virtual

Micro-segmentation bolsters security through enhanced visibility within the modern data center by:

- Increasing visibility throughout the SDDC, eliminating blind spots
- Enabling and simplifying migration to a whitelisting / least privilege / Zero Trust security model
- Providing rich contextual events and eliminating security information and event management (SIEM) false positives
- Offering inherent containment, even for Zero Day attacks

Threat analysis includes tools to complement and bolster the common perimeter logging approach. The goal of these tools is to correlate flows from different sources within a perimeter to expose threat contexts and discover compromised systems. Often these compromised systems go unnoticed for long periods of times because the suspicious traffic moves laterally inside the perimeter and does not traverse a security device – "You can't protect what you don't see."

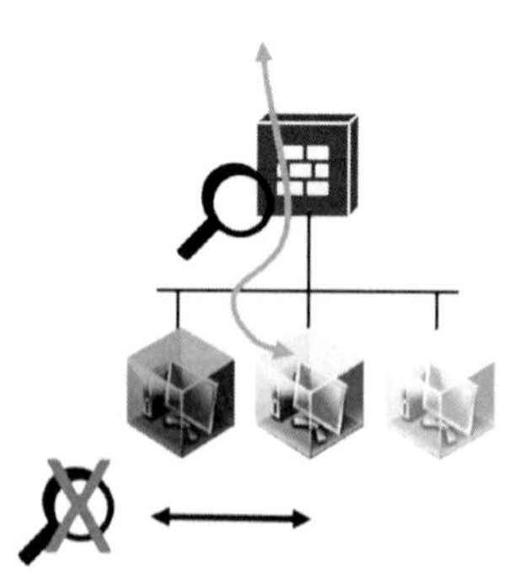

Figure 3.18 Centralized firewall blind to lateral movement

While traditional threat analysis tools at the perimeter are useful and welcome components of the security toolkit, they highlight the reality that current perimeter approaches fail to provide the proper context and visibility to identify and contain successful breaches. Are these tools being leveraged to their full potential by having them sort through ever increasing amounts of data? Could there be basic changes to the security architecture to instead provide them a lower volume of better qualified, context-rich data to help improve their efficiency?

Context

Security administrators understand the notion of context. They spend time building lists and grouping systems that have common properties: a database or PCI zone, systems that are public facing in a DMZ, users from various groups within a company, etc. In practice, they almost exclusively leverage physical segmentation and IP ranges to convey a context which can only represent one dimension at the time. For example, how best to create a VLAN, subnet or network area that would represent:

- all "Window IIS servers" plus "the ones used by MS Exchange" from
- all "Window IIS servers" plus "those used in Horizon View from
- all "Window IIS servers" plus "the ones generated dynamically for developers which need to be isolated from other users and the production network"

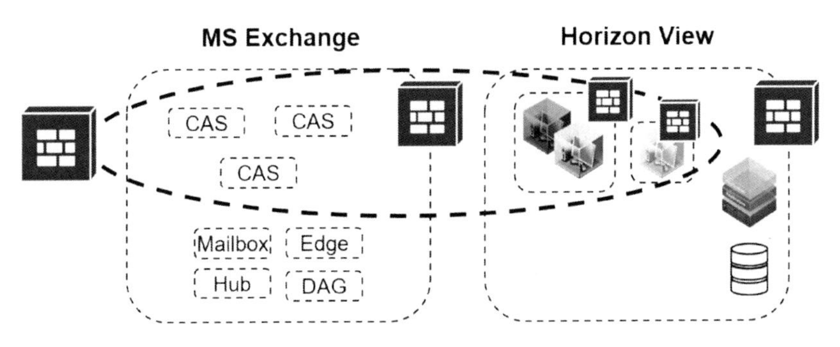

Figure 3.19 Logical security group spanning application types

A VLAN, subnet, or other form of endpoint grouping related to physical networking constructs cannot adequately represent the rich context security administrators would like to attach to applications and systems. Only a layered software construct that has no dependencies on physical infrastructure can deliver on the compound nature of the context needed.

NSX Service Composer provides a completely logical mechanism to group and arrange VMs and containers. Membership in a security group can be defined with complex logic leveraging multiple conditional statements (if / then), Boolean logic (and / or / not), and most VM/container attributes. Policies attached to the security groups will generate events based on the ruleset contained in the policies, including the identity of the user as defined in Active Directory.

The benefit of defining micro-segmented perimeters utilizing the NSX Distributed Firewall is each event sent to a SIEM is 100% representative of the full context associated with the VM or container. For example, an event could represent a blocked flow from an administrator's session on one of the IIS servers in MS Exchange located in a DMZ security group trying to access an internal system in a PCI security group. This granular level of contextual alerting is not possible with a legacy security architecture.

Visibility

To collect more information, threat analysis tools install agents on systems, collect flows from network switches facing servers, or deploy specialized probes inside perimeters. All these approaches require moving the collection engine closer to the source. They require the creation of micro-segments that have the capability to not only segregate traffic sources, but also inspect that traffic to report contextual events.

NSX provides a privileged position for security architecture via hypervisor kernel level modules. NSX native tools Application Rule Manager and Endpoint Monitoring (available in NSX for vSphere 6.3 and later) allows on the wire or within the guest application flow visibility, respectively (these tools are covered in further detail in chapter 5). By instantiating NSX Distributed Firewall on every VM and allowing partner solutions to attach at the same point through the NetX framework in addition to native capabilities, NSX provides a comprehensive micro-segmentation solution. It delivers both protection for and visibility of all traffic originating from or destined to every VM or application.

In the context of visibility, endpoint solutions, probes, and flow collectors are simply logging agents. Deploying NSX Distributed Firewall in an environment provides full visibility of virtual machine traffic in the data center by design, offsetting the expense of deploying and operating a parallel traffic monitoring environment.

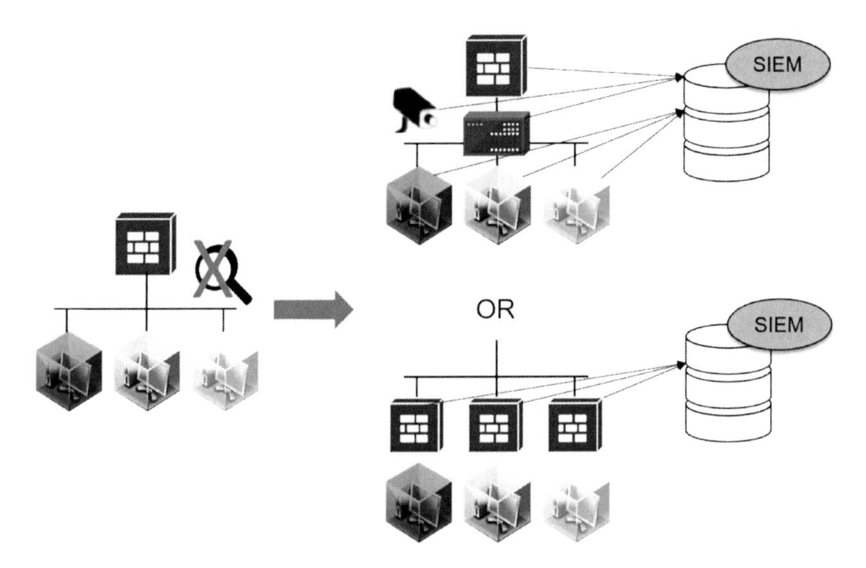

Figure 3.20 Centralized security and event loggers vs contextual visibility with micro-segmentation

Containment

Threat analysis tools alert when something goes wrong. Once the nature of an attack is understood, some tools go a step further and update security devices with new rules to contain the threat. In the interim, there will likely be multiple systems compromised and the potential for data exfiltration in an environment not micro-segmented.

It is not sufficient to simply identify the attack; it also must be contained as much as possible while the analysis is ongoing. This is where implementing micro-segmentation with a whitelisting / least privilege model becomes a critical element of the architecture. By enabling whitelisting / least privilege, only the allowed communications between known systems are permitted while other combinations are denied. Lateral movement between unrelated systems is impossible by default, making the progression of the attack throughout the data center much harder to achieve.

This model requires an understanding how specific applications work. Security administrators and application owners rarely have sufficient information to properly determine all access and communication restrictions. Often applications are deployed and security administrators simply block known bad traffic. They follow by opening a few initial ports, then iterate on the process, hoping that everything going through is good traffic.

A whitelisting / least privilege model starts by fingerprinting the application to know precisely how and where it communicates. With this list, a ruleset is built to allow only communication between specific elements, denying everything else. This approach provides knowledge of exactly what should be allowed and monitored, so a simple "deny" becomes a catch-all statement for all other possibilities.

While the benefits of transitioning to a whitelisting model are widely recognized, it may be difficult for organizations to implement due to the segmentation of legacy physical networks and distribution of applications across different security zones. Tools such as NSX Application Rule Manager, NSX Endpoint Monitoring and vRealize Network Insight address this by using VMs, objects, groupings, and network traffic to fingerprint an application to determine internal and external flows, client connections, and other essential relationships.

Once the information is gathered, a security group can be created to provide the exact context needed to create a whitelisted / least privilege policy model. This ensures only authorized traffic is flowing or being monitored, replacing the requirement to monitor all the traffic and isolate individual flows for a given application.

False Positives

A blacklisting approach, common in traditional security architectures, generates a large volume of incident data - i.e., events sent to the SIEM that might be an indication something bad is happening. Without a dedicated team sorting through these events, large amounts of this data will usually be ignored and rarely investigated as security administrators are overwhelmed due to the sheer volume. This approach trains administrators to dismiss most information because it has no relevance.

There is a need for analytic tools to sort through large amounts of incident data to find legitimate security events; however, solutions that attempt to provide visibility and context without utilizing micro-segments typically generates a deluge of uninspected data.

What if instead of sending a massive amount of data without context, these solutions could send fine-grained contextual data? This is a powerful outcome of implementing a whitelisting / least privilege approach with NSX micro-segmentation. The following example uses a VMware Horizon virtual desktop infrastructure to illustrate this concept.

Figure 3.21 shows various elements of the Horizon infrastructure organized in different security groups to provide context. A specific security policy has been applied to the Access Point Group which:

- Allows external clients to connect to the Access Point appliances
- Allows Access Points to connect to the Connection Server Group
- Allows Access Points to connect to the vDesktop Group
- Denies everything else

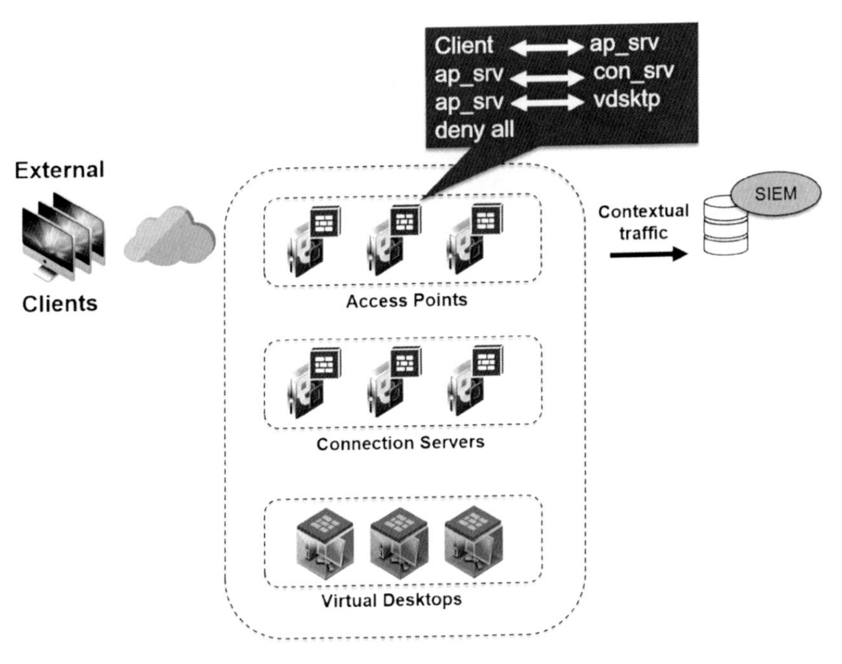

Figure 3.21 Micro-segmented VMware Horizon infrastructure

This group policy is enforced at each Access Point using NSX Distributed Firewall. It creates distinct micro-segments while also providing full visibility and context for every flow originating on or bound for any of the Security Servers.

Further exploring this example, assume one of the Access Points gets breached. The attacker or the malware will try to expand its footprint, but will realize that the IP configuration of the interface and the open ports on the system provide insufficient information to rely on for its next move, given the security policy is not based on centralized physical infrastructure or perimeters.

The deny all statement covers all the unwanted scenarios - including those that do not yet exist such as Zero Day attacks - creating a much tighter trap to contain the breach.

Any attempt to perform reconnaissance (e.g., ping, port scan, etc.) or to jump to another system other than one in the Connection Server or vDesktop groups on the appropriate ports and IP addresses, will trigger a fully qualified and 100% valid event for the SIEM while providing no information to the attacker.

A whitelisting / least privileges approach enabled by NSX micro-segmentation allows relevant incidence response data to be created rather than an overwhelming amount of uninspected data. Security administrator can investigate incidents with the assurance of not wasting efforts each time a deny event is triggered. Furthermore, monitoring and threat analysis tools can now focus on the allowed traffic for deviation and anomalies without having to sort through useless data.

NSX Micro-segmentation Redefines Security Architecture

Security professionals are struggling to keep systems and information secure due to the inherent shortcomings and compromises imposed by physical networks and security appliances. This leaves wide internal perimeters in data centers that are open for abuses should a breach ever occur. Traditional perimeter designs lack the required visibility, context, and containment to prevent company-wide breaches; thus, the current focus on probes and agents feeding threat analysis tools to compensate for the weaknesses in the architecture.

While these tools bring value and should be part of the security toolkit, there is an opportunity to leverage the unique properties offered by micro-segmentation with NSX to fix the inherent flaws found in existing security architectures. This approach presents advanced tools validated information, making them much more efficient in detecting and reacting to attackers and malware.

By applying either NSX Application Rule Manager - covered in more detail in Chapter 5 - or vRealize Network Insight to NSX micro-segmentation, it is possible to fingerprint any application and migrate from a blacklisting approach to a whitelisted / least privilege model. Working together, they provide full visibility and context for all the flows in the SDDC. This eliminates false positives, providing only qualified alerts for the security administrators to investigate while making it harder for the attack to escape the breached system and expand inside the data center.

Security administrators are not in the business of losing, and VMware NSX micro-segmentation provides an architectural shift to change the landscape to their advantage.

Chapter 4

Creating a Security Group Framework

When planning a transition from legacy security controls to agile security with micro-segmentation, a framework for security groups is needed. This framework forms the foundation for security policy and has implications on the number of security groups, security tags, DFW sections, rules and security policies needed to secure an environment.

Below is an example of a basic framework. Additional tiers may be created based on environmental needs and the required granularity. This framework will allow construction of dynamic security groups and control membership through either dynamic or static objects such as security tags, virtual machine names, and logical switches.

The security grouping framework can be split into three classifications:

- Infrastructure
- Environmental
- Application

Infrastructure Level Grouping

Workloads within an organization often share the same infrastructure characteristics. Infrastructure characteristics commonly used to form the basis of a security group include operating system type, service management level, and shared services types.

Operating System Grouping

For a basic grouping framework, it is common to see groups based on operating system types. These group can be related to specific OS versions (e.g., Windows 8, Win2012, Win2008, Win2003), OS families (e.g., Windows, Linux), or a combination of both.

Service Management Grouping

Machines that share the same operating system may not have the same service management requirements. Service management refers to the level of interaction with other infrastructure services that is required for the machine to exist within the environment prior to having any applications installed. These requirements include services like monitoring, backup, and remote management. Where one Windows 2008 machine may utilize all these services, another Windows 2008 system may operate independently of all management and monitoring.

An extra level of infrastructure level grouping may be used for service management. A common grouping strategy organizes systems based on their involvement with the management ecosystem:

- High Touch
- Medium Touch
- Low Touch

Shared Services Grouping

Every workload in an organization is required to access some form of shared service. For example, most workloads will need to access common services such as DNS and NTP.

These shared services are often pre-existing within the organization, initially existing outside of the NSX managed environment. Regardless of the service location, a reference framework must be created.

Each shared service should have a corresponding security group created. This group will include all workloads and/or IP addresses that will be providing the services. It will be used in the source or destination of a DFW rule or Service Composer security policy. This allows a single grouping methodology - regardless of the method of configuration used - to configure access to the shared service for both virtual and physical services.

Examples of specific shared service security groups are as follows:

- DNS servers
- NTP servers
- DHCP servers
- LDAP servers

When using a security group for each shared service type, adding a new DNS server simply requires adding either the VM or appropriate IP Set to the security group created for the DNS servers. No modification of the DFW rules or security policies is required.

Additionally, using individual security groups for each shared service allows fine grained access to specific shared services. These shared services can also be environmentally specific for extra granularity, as detailed in the following list:

- DNS Servers - Prod
- DNS Servers - Dev
- DNS Servers - DMZ
- NTP Servers - Prod
- NTP Servers - Dev
- NTP Servers - DMZ

The downside of this approach is security group sprawl; it is important to balance security granularity against group count.

Note: It is possible to use a single security group for a shared service across multiple environments while restricting the VMs in a single environment to individual components of that shared service. This requires specifically placed DFW rules and/or security policies. While this approach will reduce the amount of security groups, it will increase the complexity of DFW rules and/or security policies.

Environment Level Grouping

Most organizations have multiple operating environments. An operating environment can be translated into a physically or logically separated group of workloads that share the same environment characteristics. This is not the same as an OS grouping construct, as the environment may contain more than one type of operating system. One example is the differentiation of workloads between production and test/dev environments.

Environments may map directly through to the type of security posture that must be imposed on the workload (e.g., DMZ, internal, external, PCI) or directly correlate to a systems development lifecycle (e.g., production, development, test, UAT).

Examples of environment level grouping include:

- Production
- Development
- UAT
- Test
- PCI
- DMZ
- Internal/Private
- External/Public
- B2B
- eCommerce

Application Level Grouping

An application may consist of a single workload or multiple workloads, be spread across a single tier or multiple tiers, or span multiple functional groups. To aid in management, there is a need to standardize on the method of group creation for an application.

There are multiple options for grouping workloads by application. Central considerations include:

- Number of applications in the environment.
- Level of granularity required from a security standpoint.
- Level of automation required.
- Cloud management portal choice.

Entire Application Grouping

In this broad approach to application security grouping, all workloads pertaining to an application are grouped together. The pros and cons of this approach are examined in the following table (Table 4.1).

Table 4.1 Entire application grouping pros and cons

PRO	CON
Simplistic grouping structure	Loss of fine grained security enforcement within the application.
Ease of management	May allow traffic/connections to workloads within an application that are not required. For example, allowing port 443 for the web servers would also allow 443 into the DB servers.
Reduced number of security groups	

Application Tier Grouping

3-tiered applications are common in modern data centers, grouping distinct systems as front end/UI, middleware/business logic, and database/data. While organizations may collapse or create additional tiers, common tiering structure contains groups in the following areas:

- Web
- App
- DB

Creating security groups based on application tiers brings different tradeoffs than the entire application approach.

Table 4.2 Application tier grouping pros and cons

PRO	CON
Provide more granularity when configuring rules and security policies	Increase in number of security groups
Clearly defined and generalized grouping constructs	Increase in number of rules and/or security policies.
Limited number of application tier groups	May allow traffic/connections to workloads within a tier that are not required. For example, allowing port 443 into the web tier for the web servers would also allow 443 into any other server that may also sit in the web tier but is not listening on port 443.
	Increase in operational complexity due to the added layer of security grouping
	Increased complexity in deployment with a cloud management platform

Application Functional Grouping

Functional grouping helps deliver secure environments under the principle of least privilege, allowing only explicitly permitted traffic to and from workloads based on application function. This grouping strategy requires creating a distinct security group for each role within an application. Application Functional Grouping can be used to address applications that do not adhere to a standard 3-tier structure. This can include large scale applications and micro-services.

Table 4.3 Application functional grouping pros and cons

PRO	CON
Provides extreme granularity when configuring rules and security policies	Increases the number of security groups.
Adheres to the "principle of least privilege"	Increases the number of DFW rules or security policies
Caters to any number of functional roles or groupings within an application	Increases operational complexity and management

Application Grouping Considerations

Choosing an application grouping method is not an either/or exercise; it is possible to combine methods for a balance of simplicity and granularity.

Example #1

- Application = Simple Web App (SWA)
- Workloads = 8

Table 4.4 Application Grouping Example 1

Security Group	Members
SG_SWA	lswa_srv_1 swa_srv_2 swa_srv_3 swa_srv_4 swa_srv_5 swa_srv_6 swa_srv_7 swa_srv_8

Example #2

- All servers are included in a single security group for the entire application.
- Application tier groups are defined and the servers split between the tiers.

Table 4.5 Application Grouping Example 2

Security Group	Members
SG_SWA	swa_srv_1 swa_srv_2 swa_srv_3 swa_srv_4 swa_srv_5 swa_srv_6 swa_srv_7 swa_srv_8
SG_SWA_Web	swa_srv_1 swa_srv_2 swa_srv_3 swa_srv_4
SG_SWA_App	swa_srv_5 swa_srv_6
SG_SWA_DB	swa_srv_7 swa_srv_8

Example #3

- All servers are included in a single security group for the entire application.
- Application tier groups are defined and the servers split between the tiers.
- Application functional groups are configured for the Tomcat, FTP and SSO servers.

Table 4.6 Application Grouping Example 3

Security Group	Members
SG_SWA	swa_srv_1 swa_srv_2 swa_srv_3 swa_srv_4 swa_srv_5 swa_srv_6 swa_srv_7 swa_srv_8
SG_SWA_Web	swa_srv_1 swa_srv_2 swa_srv_3 swa_srv_4
SG_SWA_Web_Tomcat	swa_srv_1 swa_srv_2 swa_srv_3
SG_SWA_FTP	swa_srv_2 swa_srv_3
SG_SWA_SSO	swa_srv_3 swa_srv_4
SG_SWA_App	swa_srv_5 swa_srv_6
SG_SWA_DB	swa_srv_7 swa_srv_8

The previous examples use a flat security grouping structure. This requires a workload be a member of both the single application group and the application web tier group, manually added to each. An alternate method is to use security group nesting (described in the following section).

Table 4.7 provides a simple, fine-grained security grouping structure. Each table entry represents a security group. For an enterprise, this list can be extensive; hundreds of applications, each comprised of sets of smaller applications, all spread across multiple environments. This highlights the importance of identifying the desired level of granularity required and strategies for security consumption prior to implementing any framework.

Table 4.7 Example security group structure

Infrastructure	**Environmental**		**Application**		
OS	Environment	Service Mgmt	Application	Application Tier	Application Function
Windows 10	Production	High Touch	Active Directory	WEB	Apache
Windows 8	Dev	Med Touch	DNS Intranet	App	IE
Win2008	QA	Low Touch	vRealize	DB	SQL
Win2012	UAT		Automation	Front End	Oracle
RHEL6	PCI		Exchange	Back End	
RHEL7	VDI		VDI		
	DMZ		SAP		

Security Group Hierarchy

Security groups are the foundational building block of policy, and their structure will affect construction and implementation of the corporate firewall. A primary decision point is the choice between leveraging nested security groups (i.e., groups within groups) versus a flat structure (i.e., individual creation of all VMs/IP Sets required within the security group). Selection criteria include:

- Cloud management platform
- vRealize Automation can support a nested structure
- OpenStack and vCloud Director do not currently support nested security groups
- Management of firewall policy
- Network Introspection
- InfoSec requirements
- Monitoring

Example #4 is representative of a flat security group:

Example #4

Table 4.8 Example security group structure

Security Group	Members
SG_SWA	swa_srv_1 swa_srv_2 swa_srv_3 swa_srv_4 swa_srv_5 swa_srv_6 swa_srv_7 swa_srv_8
SG_SWA_Web	swa_srv_1 swa_srv_2 swa_srv_3 swa_srv_4
SG_SWA_Web_Tomcat	swa_srv_1 swa_srv_2 swa_srv_3
SG_SWA_FTP	swa_srv_2 swa_srv_3
SG_SWA_SSO	swa_srv_3 swa_srv_4
SG_SWA_App	swa_srv_5 swa_srv_6
SG_SWA_DB	swa_srv_7 swa_srv_8

Choosing a flat structure can be simplistic, but can add complexity to management and maintenance.

Example #5 is representative of a nested structure:

Example #5

Table 4.9 Example security group structure

Security Group	Members
SG_SWA	SG_SWA_Web SG_SWA_App SG_SWA_DB
SG_SWA_Web	SG_SWA_Web_Tomcat SG_SWA_SSO
SG_SWA_Web_Tomcat	swa_srv_1 swa_srv_2 swa_srv_3
SG_SWA_FTP	swa_srv_2 swa_srv_3
SG_SWA_SSO	swa_srv_3 swa_srv_4
SG_SWA_App	swa_srv_5 swa_srv_6
SG_SWA_DB	swa_srv_7 swa_srv_8

With a nested framework, policy can be constructed where far fewer objects are used within the firewall rule base. One caveat where Network Introspection is used in this model, currently only the top-level security group is passed to third-party firewall management platforms. If using NSX security groups on external firewalls, two options are available:

- Construct all policy used for Network Introspection with a flat structure
- Create a punt rule with all security groups within it but choose not to redirect.

Additionally, to better understand and identify all VMs and security group organization, it is recommended to use either NSX Application Rule Manager or vRealize Network Insight.

Once the framework is chosen and a methodology of consuming security groups is in place, policy creation can begin. The level of policy granularity should be tuned to meet specific organizational requirements.

Deployment Model Example

- Segmentation between environments (e.g., production, non-production, development)
- Segmentation between classifications within environments (e.g., production - PCI, DMZ, internal)
- Segmentations of tiers within classifications (e.g., web, app, DB)
- Segmentation of tier groups of like servers providing different functions of an application (e.g., for web server requiring non-standard ports, create a specific rule or policy to block non-essential services from all web servers within a security group

  Note: This would typically require a different security group to accomplish this task; if Service Composer is used, a different policy will be required)
- Segmentation of tier-to-tier traffic (e.g., web-to-web communication)

This methodology provides tunable granular control of Distributed Firewall policy built against the tradeoff of continually increasing the number of security groups and rules.

Chapter 5

Policy Creation Tools

Determining the appropriate security groups and firewall policies across numerous complex applications in an organization can be challenging. Applications, both custom and off-the-shelf, may not be well documented. This makes it hard to determine which communication paths and relevant firewall rules should be opened for the application to function while ensuring all other ports are closed to adhere to a least privilege strategy with a micro-segmentation architecture.

Gathering information about the application and its connectivity requirements by investigating its documentation or working with the application team is one way to perform the necessary application discovery. Several practices and tools exist to make this process easier.

Determining Appropriate Policies

Investigation of connection logs is a common discovery process. This consists of creating a catch-all firewall policy with a logging action, then applying that policy to the application. An investigation of firewall connection logs will allow creation of the necessary granular permit rules required for connectivity.

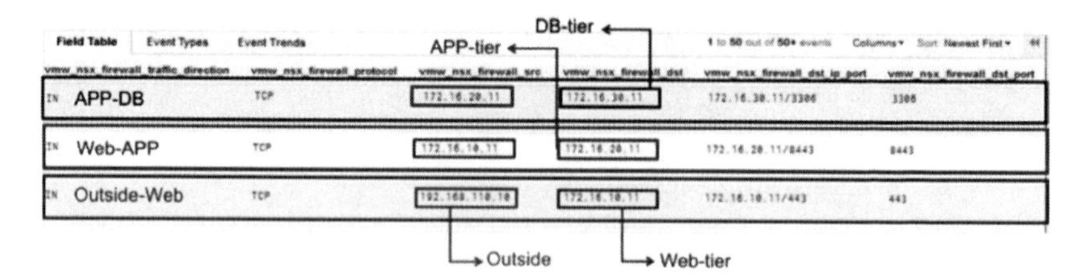

Figure 5.1 Using the vRealize Log Insight field table for application discovery

Sample Application - Intra-Application Section (Rule 1 - 3)						
1	Allow Any to Web	any	Local_Web_Tier	HTTPS	Allow	Local_Web_Tier
2	Allow Web to App	Local_Web_Tier	Local_App_Tier	TCP:8443	Allow	Local_Web_Tier Local_App_Tier
3	Allow App to DB	Local_App_Tier	Local_DB_Tier	MySQL	Allow	Local_App_Tier Local_DB_Tier

Figure 5.2 Using the vRealize Log Insight field table for application discovery

Along with the logging action in NSX DFW, VMware vRealize Log Insight can be leveraged for application discovery through connection log investigation. The use of scripting along with firewall logs aids in clean up, de-duplication, and parsing, helping to automatically generate recommended firewall policies based on the observed connections.

Another option for determining appropriate security groups and firewall policies is using vRealize Network Insight. This solution collects NetFlow data from Distributed Virtual Switches in the data center, providing network flow assessment and analytics. The network flow analytics help to determine the right security groups and firewall rules to implement a Zero Trust architecture.

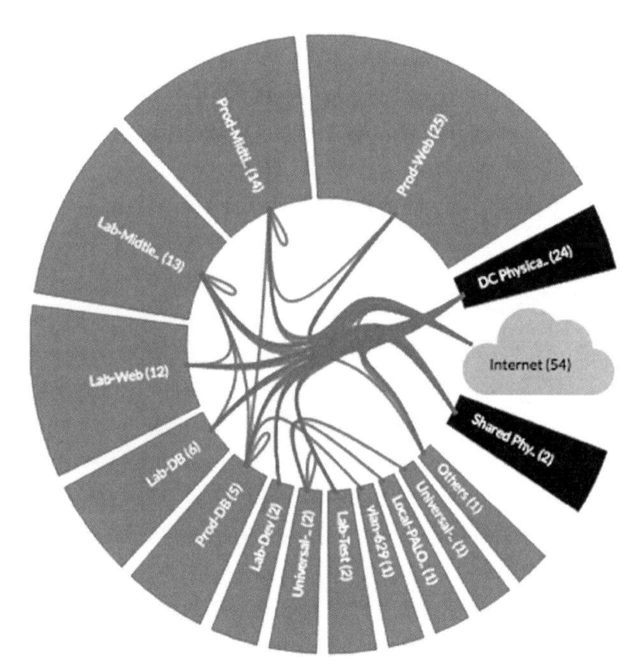

Figure 5.3 vRealize Network Insight flow analysis

The vRealize Network Insight micro-segmentation planner organizes virtual machines into logical groups based on compute and network visibility. It also provides a blueprint to put security groups and firewall rules in place. The analysis, modeling, and visualization provided by vRealize Network Insight make the process of operationalizing micro-segmentation with the right security groups and firewall rules straightforward.

With an application onboarded and secured with the appropriate networking and security controls, it is important to verify that the correct level of protection has indeed been applied. The VMware vSphere® Web Client provides visibility into both native firewall rules and third-party services that have been applied to every workload. Log analytics tools such as vRealize Log Insight with the NSX content pack or any third-party syslog collector can be used to collect logs on allowed and dropped flow, providing visibility into inter- and intra-application flows.

Another option for day 2 micro-segmentation operations is the use of vRealize Network Insight. vRealize Network Insight provides monitoring, tracking, and auditing of security group memberships and effective firewall rules, enabling rapid troubleshooting and compliance. It can generate alerts when inconsistencies occur to ensure the actual implementation remains complaint with the design.

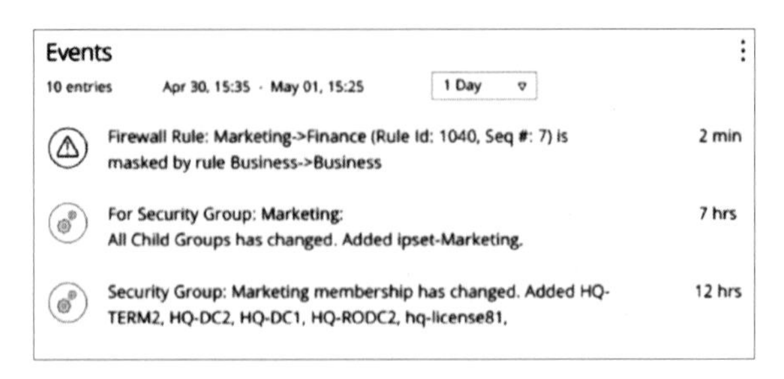

Figure 5.4 vRealize Network Insight events widget

vRealize Network Insight also provides a timeline feature, which can be used to investigate the security group membership or effective firewall policies previously applied to an application. This enables the operations team to quickly identify the cause of issues related to application functionality or compliance (e.g., an application that is no longer functioning or blocked flows between development and test environments).

NSX Visibility and Planning Tools for Micro-segmentation

NSX for vSphere 6.3 introduces two new built-in features to help operationalize micro-segmentation - Application Rule Manager and Endpoint Monitoring.

- Application Rule Manager allows a security team to profile applications on the wire.
- Endpoint Monitoring allows an application team to profile applications both within a guest and on the guest.
- Both tools can be used on a per application basis.

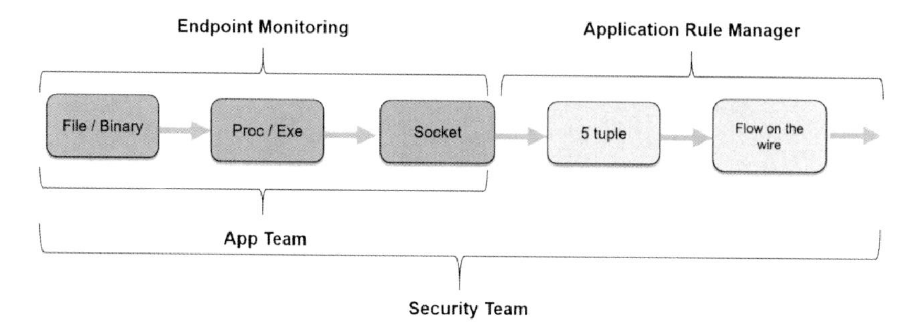

Figure 5.5 End-to-end visibility and rule creation enforcement

The benefits of these tools include:

- End-to-end visibility
- Simplified rule creation and enforcement
- App team empowerment; streamlined application deployment
- Drives whitelisting model - default deny, opening only the necessities
- Rapidly operationalizes micro-segmented applications

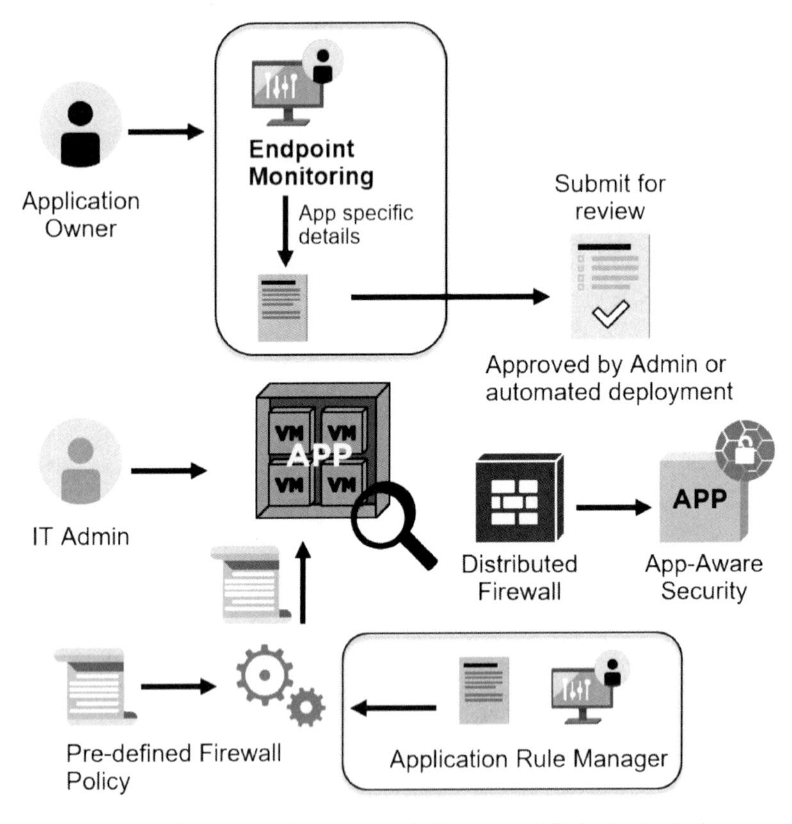

Figure 5.6 Visibility with Application Rule Manager and Endpoint Monitoring

Application Rule Manager

The NSX Application Rule Manager simplifies the process of creating security groups and whitelisting firewall rules for existing applications. Application Rule Manager can also be used for security validation and audit, providing correlated visibility between flows and recommended rules.

Application Rule Manager co-exists with the existing flow monitoring feature. Flow monitoring is used for long term data collection across the system, while application visibility is used for a targeted modeling of an application.

There are three steps in the Application Rule Manager visibility workflow:

- Select virtual machines (VMs) that form the application and need to be monitored. Once configured, all incoming and outgoing flows for a defined set of vNICs on the VMs are monitored. There can be up to five sessions collecting flows at a time.
- Stop the monitoring to generate the flow tables. The flows are analyzed to reveal the interaction between VMs. Users can filter the flows to bring the flow records to a limited working set.
- Use flow tables to create firewall rules and grouping objects.

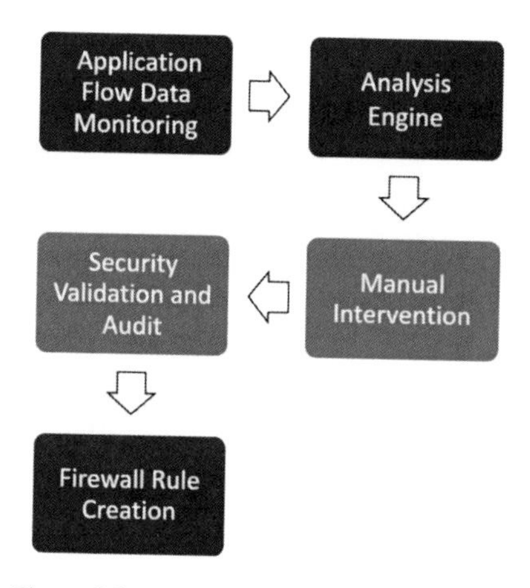

Figure 5.7 Application Rule Manager workflow

Endpoint Monitoring

Endpoint Monitoring allows an application owner to profile their application and determine the processes making network connections.

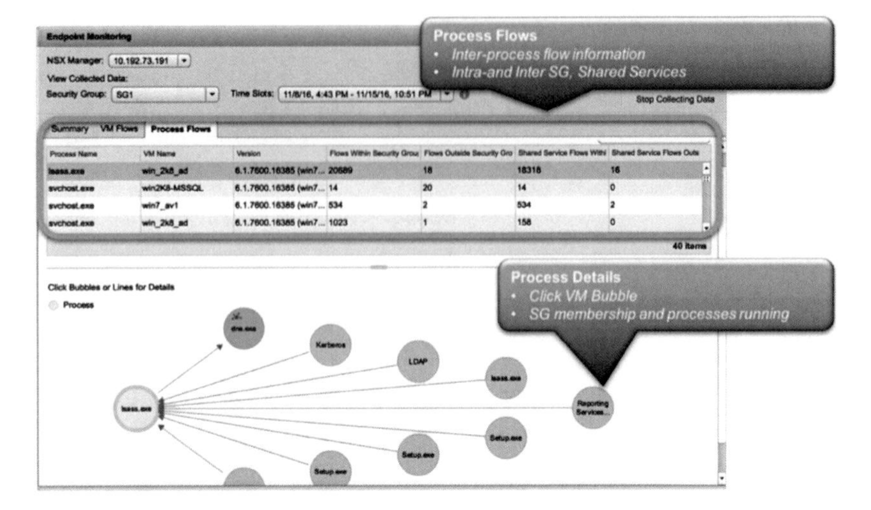

Figure 5.8 Application profiling with Endpoint Monitoring

VMs or security groups can be selected for monitoring. Activity monitoring is applied to the VMs/security groups. Requirements include VMware Tools and Guest Introspection Service VMs.

After profiling, users can see a list of:

- Details of each process running on each VM
- VM-to-VM communication
- Process-to-process communication

Endpoint Monitoring offers visual representation of the VMs, security groups, and intra-/inter-group communication, providing distinct sorting and filtering of VM and application-specific flows. These capabilities provide application owners and security administrators greater control and visibility of applications, allowing creation of more granular security controls and simplifying detection of malicious activity.

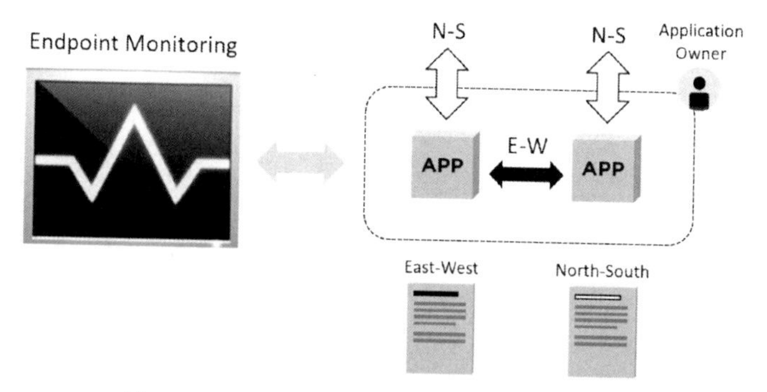

Figure 5.9 Endpoint Monitoring application visibility

vRealize Network Insight

vRealize Network Insight uses VMs, objects, groupings, and physical elements to easily fingerprint an application and determine internal and external flows, client connections, and other essential relationships.

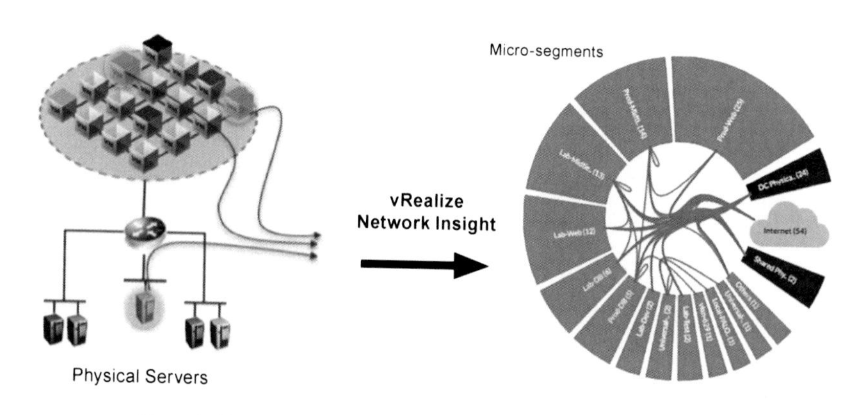

Figure 5.10 Application fingerprinting with vRealize Network Insight

Once the information is gathered, a security group can be created to provide the exact context needed to create a whitelisted / least privilege policy model and ensure only authorized traffic is flowing or being monitored. This replaces the requirement to monitor all the traffic and isolate individual flows for a given application.

Table 5.1 compares the micro-segmentation capabilities provided by vRealize Network Insight and the native capabilities provided by NSX Application Rule Manager.

Table 5.1 Visibility tools comparison

	vRealize Network Insight	**NSX Application Rule Manager**
Use Case	Network/Security Planning & Operations	Security Onboarding Using NSX Flow Records
Dataset	Using Flow Data from VDS (Pre or Post NSX implementation)	(Post NSX implementation only)
Time Period	Live & Historical Flow (default 45 days, configurable based on storage capacity)	Live Flow (~ 7 days)
Flow Collection	Always On	On-Demand
Visibility Scope	Virtual & Physical	Virtual Only

Firewall Rule Creation Using Syslog

NSX and syslog can be used as part of an incremental approach to creating, adding, and fine tuning firewall rules. In both brownfield and greenfield environments, this approach allows micro-segmentation to be implemented on a per-application basis, enabling rapid application onboarding in a secure manner.

Create Reconnaissance Rules

To begin identifying traffic within an environment, reconnaissance rules must be established. In their most basic form, these rules should allow and log all interesting traffic within an environment. These rules allow all traffic to traverse vNICs protected by DFW, logging both the initial connection and session termination - identified by the TERM statement within the syslog entry.

For example, in a very small environment, create a rule that allows any source to any destination on any service with an action of allow and log directly above the default deny rule at the bottom of the rule table. This action would enable safe evaluation of all traffic traversing the environment. Once all work is completed, the rule can be removed and traffic not matching any of the above rules will again hit the bottom deny all rule.

Best practices would limit the scope of this rule, narrowing the scope by source and/or destination or limiting the rule to specific targets (e.g., virtual machines, clusters, or security groups). It is recommended to create multiple reconnaissance rules based on the security group framework. This allows quick and efficient policy creation. vRealize Log

Insight allows quick searching of recon rules by Rule ID, helping build policy to the level of granularity required.

Where VM members of an application have been identified and assigned to a security group, the security group may be used in the applied to field of the reconnaissance rule. This will limit the results of the reconnaissance rule to vNICs that belong to the virtual machines within this security group.

An allow/log rule may be used with a source and destination of this security group. This would result in observing all traffic between members of this security group, but would not encompass traffic being sent to or from outside the group. Additional rules utilizing the group as either a source or a destination would allow for observation of traffic to/from the group.

In scoping reconnaissance rules for an application or environment, it is recommended to create a new section header prior to the creation of this rule to more easily identify where the newly created rules should logically reside. The intended use of this section header is to contain all future discrete rules that will be created for this application/ environment.

If security policies are being utilized, the discrete rules being created later will be encompassed within said policy.

Establish a Baseline and Build Policy

Once initial log data has been created via the reconnaissance rules, compare the expected results (i.e., what the application owner has provided about the application) to the logged data.

The data should be identified by the Rule ID of the reconnaissance rule. The Rule ID is automatically generated when a rule is created and will be contained within each syslog entry. Unlike rule order which can change depending on where the rule resides in the rule table, the Rule ID value is a permanent attribute of the created rule; it will not change.

When logged data confirms information provided by the application owner, create discrete rules above the reconnaissance rule to allow traffic. For example, if some hosts will have HTTP (TCP/80) access to a set virtual machines, then create a rule allowing this access above the reconnaissance rule.

When the new rule is placed above the reconnaissance rule, traffic matching the new rule will be hit first, as DFW processes the ruleset in a top down manner. The explicitly permitted traffic will no longer match against the reconnaissance rule. Focus can then shift to traffic that does not yet have explicit rules. This additionally allows identification of traffic flows that were unknown to the application owner initially.

This process can be followed iteratively until all desired traffic has identified and allowed through discreet rules.

Remove Reconnaissance Rule

Upon successfully creating all rules that allow desired traffic, the reconnaissance rule must be removed to prevent undesirable traffic from continuing to traverse the DFW.

Additional Policy Evaluation

Once an application/environment has had its policy successfully constructed, the steps should begin again for the next item. An important point to remember in the process is that the DFW processes rules like other firewalls (i.e.; from the top down).

To accurately profile traffic while ensuring that the existing rules are not inadvertently bypassed during the policy building process, it is suggested that future reconnaissance rules and their associated policies are kept below previously created rules/policies. While the risk of subverting existing rules can potentially be mitigated via choosing to target the reconnaissance rule via the applied to field, keeping reconnaissance rules below existing rules removes this risk.

Sample Steps for Rule Creation using Syslog

Note: The following steps assumes the Distributed Firewall has been enabled on the relevant vSphere clusters.

1. **Decide the default behavior for DFW (e.g., deny all or permit all for the environment);** all rule sets will follow this default behavior. A deny all rule will indicate all subsequent rules created will be opening (i.e., whitelisting) connections versus an allow all rule which indicates all subsequent rules created will be close (i.e., blacklisting) connections. This example assumes a deny all default rule is enabled.

2. **Create reconnaissance rules to allow all / deny all while also logging all traffic.** This can be done with either the DFW rule table or Service Composer.

    - If using the DFW rule table, create a section above the default section and write the rule (src/dst = any). Leave rule disabled initially. When using default deny, add the default deny rule and log rule in a section above this section, again disabling this rule.

- If using Service Composer, create a policy for the same and add the rule (src=PSG dst = any & src = any dst = PSG). Create a separate policy, but do not apply it.

  Note: In this step, no traffic will be affected in a brownfield environment.

3. **On-Boarding new applications (greenfield or brownfield) or grouping applications in brownfield.** Each application is represented as a security group. Each security group will contain a combination of virtual machines and/or IP addresses that represent physical systems. Create a security group for an application and apply the rules created in step 2.

   - For the DFW rule table, explicitly mark apply to as this security group.
   - For Service Composer, apply the security policy to a security group.
   - In NSX for vSphere 6.2, change the global setting Apply to Policy Security Group.

   Note: In this step, no traffic will be affected in a data center.

4. **Monitor Logs and use a Syslog Analyzer to determine rules.** Logs collected from the traffic will be sent to a syslog collector. vRealize Log Insight, general syslog platforms, or other third-party solutions may be used to analyze traffic patterns. Patterns will emerge that help write the appropriate rules. As applications are examined at a single point in time, the volume of rules to measure and evaluate is small.

5. **Create Shared Services rules.** Most environments will have a set of resources that all VMs need to reach, such as DNS servers, AD Servers, NTP servers, and Domain Controllers. Create a section/policy for these shared services. This will be applied above the section created in step 2 or in a separate security policy. Apply these rules and monitor the reduction in traffic hitting the previously created rules. This is residual east-west traffic that requires further investigation.

6. **Apply the default deny rule created in step 1 only to this security group.** Verify that there are no more log messages for traffic to be allowed.

7. **Continue for other applications.** On-boarding of new applications while creating rules for already onboarded applications can happen simultaneously.

Chapter 6

Conclusion

Virtualization, cloud, and software-defined services have spearheaded the modernization of the data center, upending established IT models of resource provisioning and consumption. This modernization drives the need to evolve security solutions from static, legacy models to dynamic, policy driven, granular, flexible models that can protect today's agile workloads with the necessary security controls. Micro-segmentation enables a fundamental architectural shift, making whitelist/Zero Trust Model feasible within the modern dynamic data center. NSX micro-segmentation provides the tools and capabilities needed to build a firm foundation, securing the modern data center.

Bibliography

Automation Leveraging NSX REST API Guide
https://communities.vmware.com/docs/DOC-31921

Forrester Research In Response to: NIST RFI# 130208119-3119-01 Developing a Framework to Improve Critical Infrastructure Cybersecurity
http://csrc.nist.gov/cyberframework/rfi_comments/040813_forrester_research.pdf

Micro-segmentation Defined - NSX Securing Anywhere Blog Series
https://blogs.vmware.com/networkvirtualization/2016/06/micro-segmentation-defined-nsx-securing-anywhere.html#.WIUcKrYrIUE

NIST 800-125b Secure Virtual Network Configuration for Virtual Machine (VM) Protection
http://nvlpubs.nist.gov/nistpubs/SpecialPublications/NIST.SP.800-125B.pdf

NSX and vRealize Automation Micro-Segmentation Tech Guide
https://communities.vmware.com/docs/DOC-32774

NSX-V Multi-site Options and Cross-VC NSX Design Guide
https://communities.vmware.com/docs/DOC-32552

VMware® NSX for vSphere Network Virtualization Design Guide ver 3.0
https://communities.vmware.com/docs/DOC-27683

VMware NSX for vSphere Documentation
https://www.vmware.com/support/pubs/nsx_pubs.html

VMware® NSX® Micro-segmentation Cybersecurity Benchmark
https://communities.vmware.com/docs/DOC-32904

VMware NSX Security Hardening Guide
https://communities.vmware.com/docs/DOC-28142

VMware vSphere 6 Documentation
https://www.vmware.com/support/pubs/vsphere-esxi-vcenter-server-6-pubs.html

Index

N

O

P

R

S

T